Developments in environmental biology of fishes 24

Series Editor
DAVID L.G. NOAKES

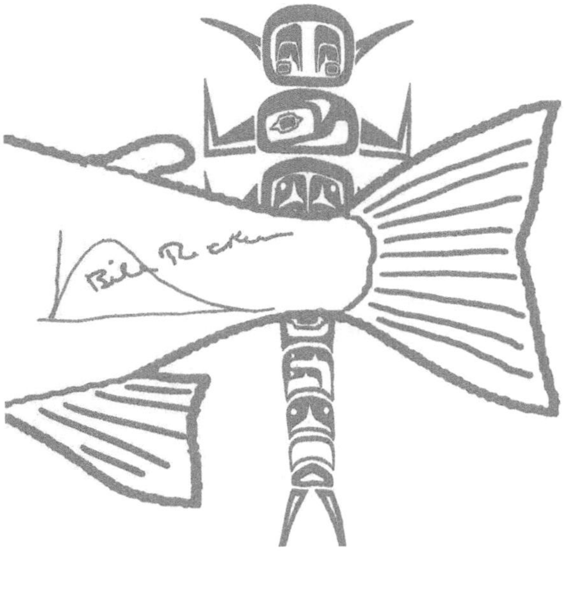

Bill Ricker: An Appreciation

Editor:

David L.G. Noakes

Reprinted from *Environmental Biology of Fishes*, Volume 75 (1) 2006

Springer

A C.I.P catalogue record for this book is available from the Library of Congress

ISBN 1-4020-4707-x

Published by Springer,
P.O. Box 17, 3300 AA Dordrecht, The Netherlands

Sold and distributed in North, Central and South America
By Springer,
101 Philip Drive, Norwell, MA 02061, USA

In all other countries, sold and distributed
By Springer,
P.O. Box 322, 3300 AH Dordrecht, The Netherlands

Printed on acid-free paper

Thanks to Stan Gregory for contributing the Plecoptera image and to Lynn Bouvier for designing the cover for this volume.

Printed in the Netherlands

CONTENTS

Bill Ricker: An Appreciation

Guest Editor: David L.G. Noakes

Frontispiece: Bill Ricker, on the day of his graduation, June 1926, from secondary school in North Bay, Ontario, Canada.

Environmental Biology of Fishes (2006) 75:1–5
DOI 10.1007/s10641-005-2440-0

Bill Ricker: a tribute

David L.G. Noakes
Fisheries and Wildlife Department, Oregon State University, Corvallis, Oregon, 97331-2802, USA (e-mail: david.noakes@oregonstate.edu)

Received 5 June 2005 Accepted 26 June 2005

Key words: Ricker's Curve, entomology, fishery science, ornithology, botany

Synopsis

William Edwin (Bill) Ricker (1908 – 2001) is best known as one of the founders of fishery science. He was also internationally recognized as an entomologist and a scientific editor. In an accompanying article written shortly before his death, Bill gives his own recollections of his career and intellectual development. A bibliography of Bill's scientific publications, translations and manuscripts has been compiled by his son Karl, and accompanies this article. Karl has also written a detailed account of Bill's early education and his accomplishments in botany and ornithology. Geoff Scudder summarizes Bill's contributions to entomology, Jon Schnute provides both personal and professional insights into Bill's mathematical accomplishments, and Dick Beamish and Don Noakes present personal recollections of Bill as a scientist and colleague.

Introduction

Bill Ricker lived a long, rich and productive life (Figure 1a, b). He contributed to many branches of science and was recognized internationally as a kind, generous and creative colleague (Beamish et al. 2003). His name will forever be associated with the Ricker Curve (Figure 2a, b), not only in fishery science but also in many other areas of ecology (Ricklefs & Miller 2000). A recent search on the Internet for his name produced more than 8000 hits, for example. The government of Canada recognized him when they named the research vessel W. E. Ricker in his honour (Figure 3a, b). The list of honours and recognitions awarded to Bill during his lifetime began with his earliest education and continued to the end of his life (Figure 4a, b). Bill did not keep a list of those awards, but Beamish & Noakes (2006) have done so in their article in this issue.

Bill was born in Waterdown, Ontario and lived the first years of his life in Guelph, Ontario where his father was a secondary school teacher. He told me the story that he was always fascinated by trains. One of his earliest memories was going down the street from his house to watch each passing train. His mother indicated a line on the sidewalk and instructed Bill not to go pass that line. I suggested to Bill that this might have been the origin of his graphical perspective on the world (Figure 5).

Bill's scientific productivity is extraordinary, for his originality and creativity, his depth, his breadth and the sheer length of his career (Garfield 1982). His first paper (Harkness & Ricker 1929) was on fish biology. He was still publishing 70 years later (Ricker & Schnute 1999), this time on astronomy! His article in this special issue of the journal extends his publication record over 75 years.

Among his many other contributions, Bill wrote articles on subjects as diverse as Sherlock Holmes (Figure 6) (Ricker 1995) and extraterrestrial fireballs (Schnute & Ricker 1999). His devotion to his family was legendary. He learned to play the double bass well enough to join his sons in the Nanaimo Symphony (Beamish & Noakes 2006). He taught himself

Figure 1. (a) The unassuming office door of Bill Ricker's office at Pacific Biological Laboratory of the Fisheries Research Board of Canada, Nanaimo, British Columbia. (b) Bill Ricker in his office at the Pacific Biological Laboratory of the Fisheries Research Board of Canada, Nanaimo, British Columbia in July 1999.

Figure 2. (a) Ricker's Curve, the entrance to the Pacific Biological Station, Hammond Bay Road, Nanaimo, British Columbia, Canada. (b) Bill's signature in my copy of the 'green book', a personalized version of the Ricker Curve.

Russian so successfully that he subsequently published a Russian–English dictionary that is still widely used (Ricker 1973, Figure 7). His facility with languages was shown in many aspects of his life, including entomology. His work on stoneflies, Family Plecoptera, was truly pioneering (Scudder 2006). It is said that his work in entomology was so impressive that everyone assumed there must be two people with the same name, one working on population biology and fishery science (Ricker 1958), and the other working on entomology. His contributions to entomology went far beyond taxonomy and systematics. His early publications on the evolution of flight, and the occurrence of flightless species of stoneflies (Ricker & Ross 1975) are well worth reading in the context of ongoing discussions on this topic (Marden & Kramer 1995, Will 1995).

I invited Bill to submit a manuscript, based on his seminar at the University of Guelph when he received his honourary D. Sc. degree. As time passed, it became increasingly obvious that much more was needed to put Bill's article in perspective. Fortunately Geoff Scudder, Jon Schnute, Dick Beamish and Don Noakes responded to my request to provide invited manuscripts on their own contacts with Bill and his work. Bill's son Karl responded with enthusiasm to my invitation, and invested a great deal of effort to document Bill's scientific activities. Karl provided extensive manuscripts on Bill's academic background, his ornithological activities, and most importantly a complete bibliography of Bill's writings.

A brief explanation is required for some details of Bill's article, and those by his son Karl, in this

(a)
(b)

Figure 3. (a) The Canadian Coast Guard Research Vessel, W. E. Ricker, docked at the Pacific Biological Station, Nanaimo, British Columbia, Canada. (b) Bill Ricker reaches iconic status on a teenager's sweatshirt.

(a)
(b)
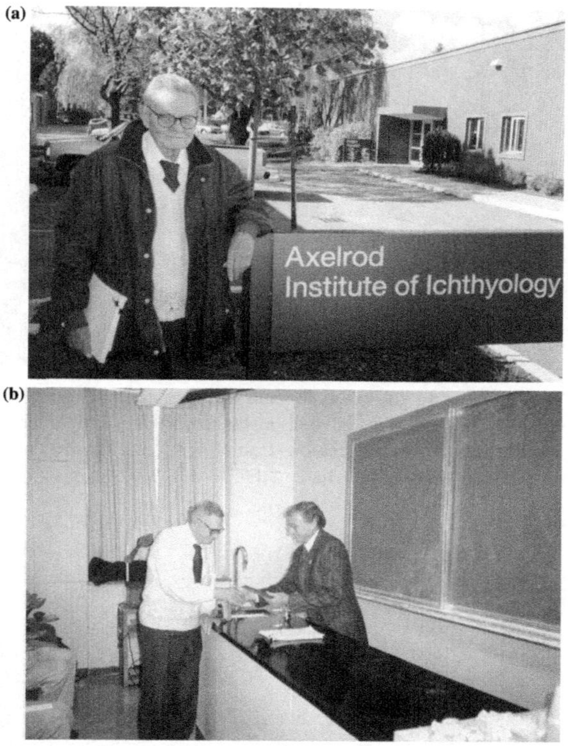

Figure 4. (a) Bill Ricker at the Axelrod Institute of Ichthyology, University of Guelph in October 1996. Bill received his honourary D. Sc. from the University of Guelph in recognition for his lifetime contributions to science. The Ricker Recruitment Laboratory in the Institute of Ichthyology was named in Bill's honour and in recognition of his pioneering contributions to fishery science. (b) Bill Ricker receiving the University of Guelph Chapter of Sigma Xi Lifetime Achievement Award from Usher Posluszny, 10 October 1996. The text from his lecture, a life among the fishes, is published in the companion article in this issue.

special issue. As much as possible I have kept all the details of Bill's article, including the captions for his original photographs, for obvious historic reasons. This includes an idiosyncratic use of numeric references to Bill's version of his lifetime list of publications, in addition to the conventional format for references cited in the text. Bill explains the details of his version of his lifetime list of publications, including his numeric system. His personal list is published exactly as he submitted it because he refers to it extensively here and elsewhere. Karl has produced a more extensive list of Bill's publications, and has attempted to cross-reference his list to Bill's. For that reason Karl's list is also published here in its entirety. Karl's article on Bill's education and predisposition to botany is published here with minimal editorial changes from his original text. This article closely parallels a number of details in Bill's own article and significantly adds to it. Karl's manuscripts were completed after Bill's death so while Karl and the other authors in this issue had the benefit of reading Bill's manuscript he did not have the opportunity to see any of the others.

Figure 5. Detail of sidewalk near railway tracks where Bill Ricker's mother might have indicated the line beyond which he should not go to view passing trains. Any resemblance to the later Ricker Curve is coincidental.

Figure 6. Sherlock Holmes Pub, on Baker Street, in Guelph, Ontario. A little known facet of Bill Ricker's career was his interest in Sherlock Holmes and his writings on the subject (see lifetime list of publications).

A great deal more could, and should, be written about Bill Ricker, his science, his personal and professional accomplishments and his influences on so many others. We still have a great deal to learn from Bill.

Figure 7. Bill's signature in my copy of his Russian–English dictionary.

Acknowledgements

I thank the members of Bill's family, especially sons Karl and Angus, for their comments, insight and suggestions. Dick Beamish, Bill Beamish and Don Noakes also provided comments on an earlier draft of this manuscript. Megan Noakes allowed me to photograph her Ricker sweatshirt. David Geber confirmed the identification of Bill's double bass.

References

Beamish, R.J. & D.J. Noakes. 2006. Bill Ricker: A man of gifted intellect, insatiable curiosity and generous spirit. Environmental Biology of Fishes. 75: 119–129.

Beamish, R.J., D.J. Noakes, D.L.G. Noakes & F.W.H. Beamish. 2003. William E. Ricker, OC, FRSC, LLD, DSc. Canadian Journal of Fisheries & Aquatic Sciences 60: iii–v.

Garfield, E. 1982. This week's citation classic: W. E. Ricker. Stock and recruitment. J. Fish. Res. Board Canada 11: 559–623, 1954. Current Contents 21: 20.

Harkness, W.J.K. & W.E. Ricker. 1929. A preliminary study of some trout waters of Ontario. Transactions of the American Fisheries Society 59: 256–265.

Marden, J.H. & M.G. Kramer. 1995. Plecopteran surface-skimming and insect flight evolution. Science 270: 1685.

Ricker, W.E. 1958. Handbook of Computations for Biological Statistics of Fish Populations. Fisheries Research Board of Canada, Bulletin 119. 300 pp.

Ricker, W.E. 1973. Russian–English Dictionary for Students of Fisheries and Aquatic Biology. Bulletin of the Fisheries Research Board of Canada 183: 428 pp.

Ricker, W.E. 1995. Moriarty, Fermat and Einstein. An addendum to "The absent corpse mystery solved". Canadian Holmes 19: 10–11.

Ricker, W.E. & J.T. Schnute. 1999. A westward trajectory extension for the earth-grazing fireballs seen on 9 February 1913. Canadian Field-Naturalist. 113: 693–697.

Ricker, W.E. & H.H. Ross. 1975. Synopsis of the Brachypterinae (Insects, Plecoptera, Taeniopterygidae). Canadian Journal of Zoology. 53: 132–153.

Ricklefs, R.E. & G.L. Miller. 2000. Ecology, 4th edition. Freeman, New York. 822 pp.

Scudder, G. 2006. Bill Ricker's entomological contributions. Environmental Biology of Fishes. 75: 111–117.

Will, K.W. 1995. Plecopteran surface-skimming and insect flight evolution. Science 270: 1684.

Environmental Biology of Fishes (2006) 75:7–37
DOI 10.1007/s10641-005-2441-z

One man's journey through the years when ecology came of age

William E. Ricker*
Pacific Biological Station, Nanaimo, V9R 5K6, British Columbia, Canada (e-mail: david.noakes@oregonstate.edu)

Received 13 April 2001 Accepted 16 April 2004

Key words: Ricker curve, fishery, statistics, translations, entomology, management, biography

Introduction

David Noakes asked me to write a summary of a talk I gave at the University of Guelph in October 1996, called 'A life among the fishes'. I agreed with some reluctance, partly because there are so many things I could have done better, or sooner, if I had been better informed or more alert at the time; but also because of the 'Doctor Watson effect', whereby everything becomes simple and obvious once it has been discovered.

The talk was a mixture of geographical settings and scientific developments in several fields. Here I will try to avoid any confusion by separating the paper into two parts. Part 1 outlines the scenic backgrounds in the order that I encountered them, and Part 2 treats the science in more or less homogeneous chunks.

My school years were spent in North Bay, Ontario. This town sits by the fairly large Lake Nipissing, whose extensive beaches shoal out into the water with 3 or 4 underwater sandbars along the way – a phenomenon for which I have not yet seen the physical explanation. Landward from the beach dunes there were gneiss outcrops covered with blueberries in August, and small lakes and streams, some of them cool enough to contain trout. This still seems to me the normal type of landscape. However, extended visits to my grandparents' home at Plattsville on Oxford County make me familiar with the rolling hills and plains of southwestern Ontario, where vesper sparrows, *Pooecetes gramineus*, sing at evening twilight and fireflies (Lampyridae) flash their signals across damp pastures.

From North Bay I went to the University of Toronto in 1926, obtaining a B.A. degree in 1930 and a M.A. in 1931. I did additional graduate work at Toronto during the spring term of 1934 and the fall term of 1935. This continued, unofficially, through that winter, when I visited about 20 universities and biological stations in Europe. These included Theinemann's Hydrobiologische Anstalt at Plon in Holstein, Rossolim's establishment beside snow-covered Lake Beloe near Moscow, and Turner's beautiful mountain laboratory at Lunz am See in Austria.

My first regular employment, starting late in 1931, was with the Biological Board (later the Fisheries Research Board) of Canada at Cultus Lake, a satellite of the Pacific Biological Station at Nanaimo, British Columbia (Figures 1, 2)[1]. In 1938, I was transferred to the International Pacific Salmon Fisheries Commission when it assumed responsibility for sockeye salmon research on the Fraser River. Early in 1939, I moved to Indiana University at Bloomington, Indiana. In 1950, I was appointed Editor of Publications for the Fisheries Research Board, stationed at the Pacific Biological Station. With a few changes of title and duties, I stayed in British Columbia until retirement in 1973, with the exception of a year in Ottawa as Acting Chairman of the Board. The

*Deceased September 2001

[1] All photographs are from personal collections of Bill Ricker, dates of photographs are given in parentheses when available.

Figure 1. Fisheries Research Board of Canada laboratory at Cultus Lake, British Columbia (1932 or 1933).

editorial and other prescribed duties of these positions were light enough that I was able to make a variety of analytical studies and reviews, and since retirement I have continued, at an increasingly slower pace, to look into problems that seem interesting.

Part 1: The setting

Lake Ontario

Before World War I, the Biological Board had operated a station on Georgian Bay, where men like B.A. Bensley, E.M. Walker, and W.A. Clemens produced papers on the fishes, dragonflies, mayflies, and other organisms of the Bay. After the War, the university people persuaded the Ontario government that similar studies on other bodies of water would be useful. An agreement was reached that additional studies be conducted cooperatively, using the name Ontario Fisheries Research Laboratory (OFRL). W.A. Clemens became its first Director. After a year on Lake Erie the work was moved to Lake Nipigon. This lake had become accessible to commercial fishing only a few years earlier when the Canadian Northern Railway reached MacDiarmid on the lake's eastern shore, so it seemed desirable to study it while its fish populations were still little altered by human activity. The several years of work there ended in 1926.

In 1927, attention shifted to Lake Ontario, the most fascinating of the Great Lakes because of the many signs of its recent postglacial history. Its outlet had tilted upward, drowning the valleys of its western rivers, which became huge cattail, *Typha* spp., swamps. At the western end a long bar cut off Hamilton Bay, duplicating a bar of the earlier Lake Iroquois that had cut off Dundas Marsh. The old Iroquois beaches mostly became railway grades that ran through, and east, of Toronto. At Scarborough the new lake was cutting back into a silt-filled channel of an interglacial river, similar to the one that became the whirlpool of the Niagara River gorge.

Lake Ontario had a long history of utilization for transport and fisheries. In 1927, its fishes were still numerous and varied, although three formerly abundant and valuable species had become extinct or very scarce: maskinonge, *Esox masquinongy*, salmon, *Salmo salar*, and sturgeon, *Acipenser fulvescens* (Figure 3). There were still active commercial gillnet fisheries at several ports, for lake trout, *Salvelinus namaycush*, whitefish, *Coregonus clupeaformis*, and ciscoes, *Coregonus artedii* (Figure 4). Also caught in deep water were burbot, *Lota lota*, a freshwater type of cod highly regarded in Europe but locally unsaleable.

After my first year at the University of Toronto I was offered a summer job as a general handyman with the OFRL, mainly at Port Credit where there was a commercial fishery. J.R. Dymond was leader of the group, which included A.L. (Andy) Pritchard who was studying ciscoes (Figure 5), while Howard Dignan and I did whatever was needful (Figure 6). At the other end of the lake, graduate student John Hart was studying the whitefish of the Bay of Quinte, and Don Rawson was working on Lake Simcoe. Both of them visited us during the summer. Our laboratory was in the loft of a

Figure 2. Water and plankton sampling on frozen Cultus Lake (plankton net suspended). Laboratory assistant is unidentified (March 1937).

Figure 3. Bill Ricker with a small lake sturgeon, *A. fulvescens*, at Lou Joyce's fish house at Port Credit, Ontario.

fish house owned by Lou Joyce. Accommodation was in tents set up in a small apple orchard belonging to the local bailiff, whose wife supplied the meals. Specimens were obtained from an assortment of small-meshed gillnets attached to one end of a long string of commercial nets set by the regular fishermen in various parts of the lake, and also from our seining expeditions up tributary rivers and along shore as far as Hamilton Bay (Figure 7). Between collecting trips, Dymond got me to measure body parts and count fin rays on a few kinds of fish. I remember the deepwater sculpin, *Myoxocephalus quadricornis*, an arctic marine fish that became adapted to brackish and fresh waters. It had been pushed south in a series of proglacial lakes and left 'stranded' in their cold bottom water when the glaciers retreated.

The account of the Lake Ontario fishes by Dymond et al. (1929) tells about the diversity and abundance of species that still existed at that time, including eels, *Anguilla rostrata*, that came up from the ocean through the St. Lawrence River. We found lampreys, *Petromyzon marinus*, spawning in the Credit River, as A.F. Coventry had in the Don River a few years earlier. Some of the lake trout, *S. namaycush*, had fresh or healed scars from their attacks, but some sort of mutual

Figure 4. (a) Gill nets drying at Joyce commercial fishery, Port Credit, Ontario (1927). (b) Gill net just lifted with two whitefish, *C. clupeaformis*, Port Credit, Ontario (1927). (c) Andy Pritchard (left) with commercial fisherman (right) setting nets on Lake Ontario, from Port Credit, Ontario (summer 1928).

accommodation had evidently been achieved. This did not exist in the upper lakes when lampreys reached them through the Welland Canal during the 1930s. Since then there have been many additional changes. I understand that the deepwater sculpins have almost disappeared, smelt, *Osmerus mordax*, have replaced alewives, *Alosa pseudoharengus*, as a major food for larger

Figure 5. Bill Ricker, Andy Pritchard, J.R. Dymond, John Hart and Howard Dignan (left to right) at Port Credit, Ontario (192?).

fish, and the steelhead trout, *Oncorhynchus mykiss*, and coho salmon, *Oncorhynchus kisutch*, are now major sport fishes.

Waters of the Niagara cuesta

A stratum of massive dolomite in the Silurian rocks of New York and Ontario has resisted erosion much better than the shales and broken limestone layers that underlie it, thus forming an escarpment that runs across the province from Queenston to Manitoulin Island. The highlands behind the cuesta up near Georgian Bay are a region of heavy snowfall because the prevailing winds arrive laden with moisture from the open waters of Lake Superior and Huron. This accumulates in swamps and cool streams that eventually fall over the escarpment at several points, forming excellent habitat for brook trout, *S. fontinalis*, both above and below the waterfalls (Figure 8).

W.J.K. Harkness had succeeded W.A. Clemens as head of the OFRL, and in 1928 he decided to start work on two major game fishes. This was at least partly in response to interest stemming from various members of the Toronto Anglers Association. A.L. (Al) Tester went up to Georgian Bay to study smallmouth bass, *Micropterus dolomieu*, while F.P. (Fred) Ide and I were to study brook or speckled trout, *S. fontinalis*, and the waters of the cuesta region of southern Ontario. At that time this trout (really a charr) was the only salmonid in those waters, and almost the only fish in the cooler ones.

Fred and I set up camp at Horning's Mills, on the banks of the millpond, which still fed an active gristmill with a high-head turbine. This small impoundment was fed by spring streams and had an extensive growth of the limy alga *Chara*, which in turn supported numerous caddis fly larvae (Trichoptera), and the trout that fed on them. We took the temperatures and the oxygen and carbonate contents of the pond and local streams, and collected aquatic fauna, especially insects, both in the water and above it. Fred had collected insects for the National Collection in Ottawa, and showed me the tricks of the operation. I collected as many trout stomach as possible from anglers who were encouraged by my offer to clean their catch.

Transportation to the various streams and ponds of the region was by means of a box-shaped model-T Ford 2-door sedan, which I bought in Plattsville for $25. It was made about 1922, and hence was started by a crank, and of course flat tires were a frequent event (Figure 9). One thing I hadn't really expected. Gasoline went to the engine by gravity from a tank under the front seat, which made it necessary to stop and blow it into the carburetor once or twice when going up any really long hill, or else go up in reverse. With this vehicle we visited several of the rivers along the cuesta. The larger ones had cut steep-walled valleys into the escarpment, with a waterfall at the head that tumbled over the hard dolomite. There was also one small natural lake and several series of trout ponds owned or leased by angling enthusiasts.

In 1930 it was arranged that Fred and I would make a more detailed study of a section of one stream, the Mad River, above the milldam at

Figure 6. (a) Off to set gill nets, Port Credit, Ontario (1927). (b) On board a Port Credit gill-netter. 'I'm not really sick, just watching the fish come up in the net' (1927).

Singhampton. This slow stream had numerous spring tributaries and beds of *Chara*, and had long been famous for its trout fishing. In addition to the summer work we made trips to it in fall and spring, as well as during one winter. With so much spring water entering it at 7° C, the Mad did not freeze completely, although parts of it we could traverse on our skis. This sortie was by train, and for the return trip Fred and I may have been the first to ski down the escarpment to Collingwood.

Lake Nipissing

In 1929 it was decided that the OFRL should study Lake Nipissing. This was a soft-water lake about 40 km long. At one stage of deglaciation it had been part of the main drainage channel from the upper lakes to the Ottawa and St. Lawrence Rivers. The headquarters chosen was an abandoned sawmill site in Frank's Bay, on the south side of the lake where it starts to contract into the

Figure 7. A.L. Pritchard and J.R. Dymond collecting fishes in the Credit River (28 June 1927).

French River (Figures 10, 11). One two-storey building remained standing, which had been a boarding house, now with a leaky roof and without windows, but with a functional brick chimney. This provided laboratory space, a kitchen and some sleeping accommodation, although bugs

Figure 8. Nottawasaga Falls where the Noisy River goes over the Niagara cuesta (21 July 1927).

from a bat colony in the attic, which greatly resemble bed bugs, caused some consternation for a while. The roof was patched with tarpaper and the windows covered with cheesecloth to keep out the blackflies (Simuliidae), which were very numerous that year. I preferred to sleep in a tent pitched on the adjacent hillside, where a whip-poor-will, *Caprimulgus vociferus*, called in the evening and pursued cockroaches (Blattidae) across the glacially smoothed gneiss.

Work proceeded much as on Lake Ontario, except that we set and lifted the gillnets by hand. After clearing them of fish, the nets were reeled up to dry, and later untangled and new mesh woven into the larger holes. Among those present were Mr. and Mrs. Harkness, and Dr. Lucas, a pharmacologist who did the chemical analyses of water samples. Beside me, there were two other 'veterans' of the previous year's work, Fred Ide and Al Tester. In 1928, Al had used a 5-m row-boat with a small Evinrude outboard motor at Georgian Bay. This he brought up to Nipissing by way of the French River, along with Murray Fallis and J.P. (Jack) Oughton – an exciting and memorable 3-day passage. Another new recruit was Ernest Pentland from Queen's University in Kingston.

Our main means of transportation on Lake Nipissing was a homemade flat-bottomed boat about 6 m long, fitted with an engine from a model-T Ford which had to be fed oil in the gasoline as well as the crankcase because the splash didn't get up to the front cylinder, but it served very well. Most of the lake was shallow enough and exposed enough

Figure 9. (a) Car used for fieldwork (1928). (b) Bill Ricker preparing the car used for research trips for winter storage at Hornings Mills (September 1928).

that the water maintained complete vertical circulation throughout the summer, providing warm aerated habitat for millions of large mayflies, *Hexagenia* spp. These were the principal food for the abundant walleyes, *Stizostedion vitreum*, there called pickerel, for which the lake was famous, along with whitefish, sturgeon, and other species. Ciscoes were present, but during the summer were crowded into a deep hole off Frank's Bay whose thermally stratified water provided a cool refuge. Our first set of gillnets happened to be in this exact spot, and produced a greater flood of fish than could be handled in 2 days of hard work. I returned to Frank's Bay for a few weeks in 1931, but did not contribute directly to the various reports on the Nipissing fauna. Notable among these was Fry's (1937) report on the ciscoes.

Algonquin Park

After 2 months of work at Lake Nipissing in 1929, it was arranged that Fred Ide and I should go to Algonquin Park for additional study of trout habitats. There was at that time no road into the Park, which indeed was a park only in that it was closed to settlers and hunters. The Canadian National Railway had a line through it from Scotia Junction to the Ottawa River, and a number of boys' and girls' camps had been built on lakes not far from it. Canoes were the established means of travel through the park, and we had a rather small birch canoe that had been made by a Nipissing Indian. Yellow and black fire hazard signs marked portages in the park. We got off the train at Joe Lake, bought some food, talked to a few people to

Figure 10. Abandoned bunkhouse of a sawmill at Frank's Bay, Lake Nipissing, Ontario. Back row: Ernest Pentland, E. McWalker, Fred Ide, Bill Ricker, Jack Oughton. Front row: Mrs. Harkness, Mrs. Lucas, D. Lucas, Lucas daughter (1929).

get the lay of the land and the water, and then set out southward. We crossed Joe Lake and then Canoe Lake, where Tom Tomson had stayed when he made some of his best-known paintings, and where his death by drowning is still a source of controversy. From Canoe we portaged to Smoke Lake and paddled down to its southern end, where we set up camp by the portage to Ragged Lake.

Logging was the major occupation in the park, but logging trucks had yet to be invented, so the work was all done in winter. Logs were hauled on sleights along narrow roads and dumped onto the ice of a convenient lake or river. In spring they were floated downstream with the flood, usually assisted by water released from low storage dams, while other dams and chutes got the logs over falls and difficult rapids (Figure 12).

During the next few weeks Fred and I worked much as we had at Horning's Mills, but using the canoe instead of the old Ford to get around. Visits were made to adjacent lakes and streams, taking temperatures, pH and oxygen contents, collecting insects and other aquatic organisms, and getting trout stomachs from the (few) anglers that we encountered. These were mostly from lake trout, whose food was predominantly age-0 perch, *Perca flavescens.* When travelling, we sometimes fished ourselves using trolling gear, but found this a very slow method of obtaining specimens, only once did we obtain a good series of *S. fontinalis,* from Wolf Lake just outside the Park boundary. This was distant from the main canoe route by two fairly long portages and was almost never visited in summer, although there was a cabin used by deer hunters in autumn. Here brook trout were abundant and hungry, perhaps because much of their food was diverted to the thorny-headed worms (Acanthocephala) that filled their intestines.

Toward the end of August it was time to leave, and Fred and I decided to traverse the Park northward to meet the CNR's transcontinental line at Brent. This was a 3-day trip from the southern line, but our map showed that portages existed between a series of lakes and on down the Nipissing River (here pronounced Nipisang). We found a pair of Forestry lookout men on one of the lakes, who said no one else had come by that year. They wanted us to stay a while, but we felt we had to get along after an hour or so. The high spots of the trip were two tracts of forest that had never been logged. One consisted of big maples, *Acer* spp., and other hardwoods, with a few very large old white pines, *Pinus strobus* that had preceded them on the site. The other, a long stretch of pines along the Nipissing River, said to

Figure 11. (a) Fred Ide on island near Frank's Bay, Lake Nipissing, Ontario (1929). (b) Teron, Harkness, Oughton, Ide, Fallis (left to right) at Frank's Bay, Lake Nipissing, Ontario (1929). (c) W.J.K. Harkness at Frank's Bay, Lake Nipissing, Ontario (1929). (d) J.R. Dymond at Ontario Fisheries Research Laboratory, Frank's Bay, Lake Nipissing, Ontario (summer 1929).

have sprung up following a fire about 1650. At our last camp there was a small pack of wolves of a reddish colour down by the river in the morning.

At Brent we had a long wait for a slow mixed train to North Bay. Leaving most of our gear there, we went south to Toronto, where there was a meeting of the American Fisheries Society at which a good representation of locals was desired. In fact, I gave my first oral presentation of the results of research – on trout foods, of course. Afterwards, I returned alone to the Park for another week, mainly in order to get additional trout from Wolf Lake.

Figure 12. Dam in Algonquin Park, Ontario used for driving logs in spring.

Cultus Lake, British Columbia

Cultus Lake, about 5 km long and 6.4 km^2 in area, takes its name from a Chinook word meaning useless, or even sinister. What was thought to be wrong with the lake I do not know, but actually it has always been very useful as a refuge from the mosquitoes (Culicidae) that infest the lower Fraser Valley, and were hellishly abundant before a large floodplain lake was drained. What keeps the lakeshore free of these pests is a brisk southwestern wind that blows daily during summer from about noon onward.

The lake lies between two mountain ridges that form the western boundary of the Cascade Range in Canada. Its basin had been enlarged by a tongue of the great Fraser Valley glacier, and when the tongue melted it left a lake that abutted on the main glacier to the north and drained southward into the Nooksack River in Washington. This lasted long enough to build up a well-defined beach. When the Fraser Valley was finally freed of ice the lake's level dropped, and for a time it became an arm of the sea during the late post-glacial marine incursion. When the land rose again the lake was held by a low moraine and drained northward into the Fraser.

Cultus Lake's fish fauna included a population of anadromous sockeye, *O. nerka*, the most valuable of the fiver western species of salmon. Starting in 1923, the Biological Board supported a study of the freshwater ecology of sockeye by R.E. Foerster, and Cultus was chosen as the principal site because of its convenient size and fairly easy access (Figure 13). A first report was published in 1925, but the study continued, partly because an experiment had been mounted to check the efficiency of the salmon hatchery operations. Late in 1931 I was employed to continue biological and other work at the lake, and Foerster moved to Nanaimo in 1932.

Mountains were new to me, of course. Fortunately there was a trail up to a small cirque lake and surrounding flowery meadows at about 1500 m elevation. This and other local climbs developed my leg muscles, as canoe journeys had developed my arms, past the point where an all-day foray was painful.

The Fraser River

The Fraser and the Columbia are the principal drainage channels of southern British Columbia, but the Fraser is one whose watershed lies almost wholly within the province. Although the entire region was covered by ice during the glaciations, the Fraser Valley contained a mainstream glacier only from Hope westward. From about Soda Creek south, the river flows in a narrow valley that can be considered one long canyon. However, the canyon par excellence is the stretch from Boston Bar to Yale, and even within that interval two especially narrow sections have been distinguished. One is the 'Little Canyon' near Yale, into which Lady Franklin and Sophia Cracroft were taken in 1861 as passengers in a large dugout canoe – a sightseeing trip arranged by the Hudson's Bay Company. The 'Big Canyon', a few kilometers below Boston Bar,

Figure 13. (a) First cabin at Cultus Lake, British Columbia (1933). (b) Boat used for studies on Cultus Lake, British Columbia (1932–1936). (c) Net reel at Smith's Falls, Cultus Lake, British Columbia (1932–1934).

has its narrowest passage at Hell's Gate, where the river pours between almost vertical granitic walls. The difference in level between high water in June and the winter low may be more than 30 m but even it its lowest there is still about 25 m of water above the bottom of the channel.

In spite of this violence, nowadays large inflated rafts take thrill seekers down through Hell's Gate, and during the gold rush a disgusted miner, in a hurry to get home, lashed himself to a large log and was carried downriver safely. What seems less believable, although true, is that a steamboat made its way upstream through the Big Canyon in 1882, using lines from its capstan and a steam winch that ran to eyebolts in the canyon walls, as well as the tug of 150 labourers pulling a third line from shore. Both the miner and the ship made the trip in early autumn when turbulence had subsided considerably.

During the summer of 1938 I was one of a party that included Earle Foerster, Gordon Thompson and Ernie Kennedy, who tagged sockeye at Hell's Gate. There are two slots that have been weathered into softer rock just below the Gate proper, forming eddies in which the salmon rested before or after trying to make the ascent. We dipped these out, measured them, and attached a round tag through the back. We had also tried using a short piece of gillnet to catch them, but found this less efficient than the dipping procedure, which had been used by the native peoples for many generations. But we were not the only ones using that approach to the river. At that time the railway had a flag station named Gorge close to the Gate, with a house by the track for a resident Section Man. The latter had a young son who had prospected down into the slot under an overhanging boulder. He found a layer of dense course gravel that none of the miners had located. Using a small sluice box he could wash out visible specks of gold.

While working at the Gate we had the roar of the rapids continuously in our ears, but the more pervasive sound was an even 'swish' that I took to be the suspended sand scouring against the granite walls. Sockeye were not the only animals making their way upriver. We also caught a few coho and Chinook salmon, *O. kisutch* and *O. tshawytscha*, and observed lampreys, *Entosphenus tridentatus*, making short spurts from one wall to the next, where they attached themselves by the sucker around their mouths.

Indiana

Indiana is the smallest of the Midwestern United States, but a very interesting one. Most of it had been covered by broadleaf forest, but there were 'islands' of tall grass prairie scattered over its northern half. It had never been completely covered by ice, for there is a narrow strip of unglaciated territory running north from the Ohio River, on which Indiana University is located. This also contains the massive beds of clear limestone that now ornament thousands of buildings throughout the continent. The forest contained many trees that were new to me, and also new birds, flowers, insects, frogs, salamanders, snakes, and so on. In fact it was the third major biome in which I worked. Another feature was an abundance of caves, a few with an active stream that contained blind crustaceans (amphipods, isopods, and crayfish), and even blind fishes (Amblyopsidae).

The Wisconsin glaciation had extended about half way across the State, and in the northeastern portion left an interlobate moraine with low hills and numerous small lakes. These were now in various stages of eutrophication or senescence, but the larger ones had relict populations of ciscoes in the cool strata below the thermocline. During the 1920s, under the leadership of Will Scott of Indiana University, a research program in fisheries-related topics was developed in cooperation with the State's Department of Conservation, called the Lake and Stream Survey. Its headquarters were at Winona Lake, where the university had a field station originally founded by Carl Eigenmann in the last century. It was the site of very early work on variability in fish populations, as well as faunistic studies. An unusual project was that of mapping the original postglacial lake basins by Ira Wilson, using a hand-operated but very effective hydraulic system to bore through the accumulated sediments.

Will Scott died in 1936, but Fernandus Payne, a Dean at Indiana University, had consulted the Conservation people and it was agreed that the work on fish, fisheries and aquatic organisms should continue. For this, I was appointed to the staff of the Department of Zoology of the university early in 1939. There I taught limnology, ornithology and, later, elementary statistics. For research projects, I was interested in fishes mainly, but the fishes of Indiana had been studied by David Starr Jordan and other experts, so I felt that it would be profitable to make intensive studies of the populations of a few lakes, especially how they were affected by fishing. We employed students to census the anglers' catches and collect scales for growth studies. The most numerous game fish was bluegill sunfish, *Lepomis macrochirus*, and these could be captured in simple wire traps that we used to obtain specimens for marking, release, and subsequent capture.

The Lake and Stream Survey also sponsored work of other types, and in other places. For example, there were Shelby Gerking's new survey of the fishes of the state with special attention to streams and rivers, and Karl Lagler's study of a bayou pond full of armour-plated gar pike (two species of *Lepisosteus*), near the Wabash–Ohio junction. Donald Scott (later at University of Georgia) studied the fishes of a woodland section of the Tippecanoe River. Other students during my time included Donald Wohlschlag, who went to the Port Aransas biological station on the Gulf of Mexico, and John Gottschalk, who rose to become head of the U.S. Fish and Wildlife Service.

Departure Bay

This bay near Nanaimo on Vancouver Island was named by a man called Pemberton, but he said nothing about who had departed or when, or where they were going. At the head of the bay are Indian village sites and middens, so evidently there were many arrivals and departures of dugout canoes in centuries past. Even today racing canoes come and go during the spring. In the late 1700s, Spanish explorers frequented these waters, and left behind names like Gabriola and Malaspina, in the late 1800s the bay was a busy port where coal was loaded onto square-rigged ships from mines near Wellington 3 km inland; the dock sites can still be seen, marked by the rock ballast that the ships discarded, these ships, I'm told, departed for ports from San Francisco to Shanghai. In 1900 the Vicar of the Anglican church in Wellington was G.W. Taylor, a man who was bursting with energy that he applied to matters both sacerdotal and secular. What concerns us is that he was interested in several branches of natural history and collected assiduously, especially insects and molluscs. For these activities he was elected a Fellow of the Royal Society of Canada, and from this vantage point he lobbied assiduously for the establishment of a marine station where researchers could come and study the fishes and largely unknown, forms of life found in local waters. His efforts, and those of other like-minded people, came to fruition in 1908, when a laboratory building was built on the north shore of Departure Bay, with a dock and small motor vessel out front. The building included accommodation for a Director, but others lived in tents on the hillside behind. This was no hardship, of course, because at first nearly all of the work was done in summer, by volunteers, from universities, colleges and high schools across the country and even abroad. There was a hope that this would benefit the local fisheries, and indeed it did provide information on the size and age structure of populations of herring, *Clupea pallasi*, salmon and halibut, *Hippoglossus stenolepis*, that are still valuable. By the time I arrived in British Columbia, the Station had acquired a residence building with kitchen facilities, although the tenters did not disappear completely until the 1960s. In 1923 W.A. Clemens was appointed Director of the Station, and starting in late 1920s a few full-time researchers were hired. Late in 1931 I became one of these. Although living at the satellite site on the mainland, I made frequent visits to the Station to do chemical analyses and to use its library.

Starting July 1, 1950, I made the Station my headquarters for work as Editor of the Fisheries Research Board's publications, and for such other studies as seemed useful, interesting, and compatible with office existence. Unfortunately I never engaged in fieldwork on salt water, but from a small boat during recreational time was able to become acquainted with the varied life of the sea. At the lowest tides in summer there was exposed an amazing variety of crabs, starfish, anemones, barnacles, chitons, snails, nudibranchs, corals, bryozoans, and brachiopods. Swimming in the water above, were small jellyfish and the nudibranch *Melibe*. At night there were luminous comb-jellies, *Ctenophora*, which with the protozoan *Noctiluca* filled the water with phosphorescence in the wake of a boat or a swimmer. Outside the bay were beds of oysters and several kinds of clams, and it was not difficult to catch an occasional coho or chinook salmon, lingcod, *Ophiodon elongatus*, rockfish, *Sebastes* spp., and so on.

Part 2: The Science

General natural history

My early observations on birds at North Bay, together with those of Doug Clarke at Frank's Bay on the other side of Lake Nipissing, were published by the Royal Ontario Museum (33). Collecting done during the 1930s in the vicinity of Cultus Lake turned up 4 dragonflies (Odonata) new to Canada (27), and about 35 each of new stoneflies (Plecoptera) and caddis flies (Trichoptera), many of them new to science. Another animal new to Canada was the ribbed frog, *Ascaphus truei*, of mountain streams, the only new-world member of its group. This frog's tadpole has a big sucker around its mouth by which it clings to stones in rapid water (11).

An interesting find, made with Ferris Neave on a small island in Kennedy Lake, Vancouver Island, was a nesting colony of a dozen or so pairs of mew gulls, *Larus canus*, the first known in America south of the boreal region (117).

The fence counts of smolt and adult sockeye at Cultus Lake and the gillnetting in the lake itself produced a lot of information about the lake's salmon, especially sockeye and coho. One surprise was to find fairly large numbers of coho living in the lake until early summer of their third year (91). Another was to find 'residual' sockeye maturing in the lake at ages 2, 3 or 4, without going to sea, with an interesting pattern of sex ratios and growth rates at the three ages (3). Reading Fisher's '*Genetical Theory of Natural Selection*' prepared me to postulate rapid evolution of wholly lake-living populations (kokanee) from these residuals, under favourable conditions (35). For a time, I thought Cultus Lake itself has a small population of kokanees, but these proved to be escapees from a shipment of fertilized eggs from Kootenay Lake (35, 107). Information on the various fishes of the lake is included in the predator control papers (36, 38), but a stock of tiny coast range sculpins, *Cottus aleuticus*, living in deep water got separate treatment (109). The persistence and exactness of the 'homing' behaviour of adult sockeye was demonstrated by an impromptu experiment in the river below Cultus Lake (12).

Trout and trout waters of Ontario

While still at university, I wrote reports on the work done each summer from 1928 onward, while the 1930 study of the Mad River served as a basis for my M.A. thesis. The principal publications concerned the trout themselves (4), the lakes and ponds they lived in (5), and descriptions and a classification of the creeks and rivers we had studied, both with and without trout (8). The main contrast was between the clear headwater streams of the cuesta region south of Georgian Bay and the brown waters of Algonquin Park, which were heavily populated by blackflies. Subdivisions were made on the basis of size, temperature, type of bottom, and so on. The insects and other animals inhabiting these waters were not then well enough known to classify them into anything but very general categories. Today, identifications to species are possible in most cases, and have been published for stoneflies (201), mayflies, caddis, blackflies, dragonflies, and other groups.

Gregory Clark, a well-known Canadian newspaperman and keen angler, had once rented fishing rights on the Mad River and caught a 2.2-kg brook trout there. He and I collaborated on a brochure for public distribution, one of a series put out by the Ontario Department of Lands and Forests.

Classification and distribution of stoneflies

In the course of the Ontario work with Fred Ide in 1928–1930, we collected aquatic insects of all sorts, except for the difficult midges and mosquitoes. Fred was making a special study of mayflies (Ephemeroptera) and E.M. Walker, a Professor at the University of Toronto, suggested that I undertake identifications of stoneflies (Plecoptera). We both, of course, collected dragonflies (Odonata), which were one of Dr. Walker's specialties. I did not get around to publishing anything on stonefly systematics until 1935. During my European tour of 1935–1936 I visited museums that contained type specimens from North America (28). Later I planned a series of papers on the stoneflies of the various parts of Canada. These have now all been published, but usually as part of a larger work by two or more authors (45, 49, 66, 72, 140, 171, 201, 206). Review of stonefly distribution in Canada was prepared for the conference in Abisko (132).

While in Indiana, I produced a paper on the local stonefly fauna (57), and a revision of some aspects of the classification of Plecoptera (87, 108), but most stonefly work there was with the excellent collection of the Illinois Natural History Survey, which H.H. Ross put at my disposal after T.H. Frison died. This served as a basis for a revision of the American species of *Paragnetina* (82), and for a major work that included descriptions of many new species and genera, and a revision of the American Perlodidae and Nemouridae (88).

After leaving Indiana I continued to study specimens sent to me for identification, and also those in museum collections, including some from the University of Guelph. From this came new records and descriptions (98, 133), a review of the genus *Isocapnia* (106), a contribution to a five-author Stoneflies of Minnesota (155, 170), and keys to the North American genera (108). About 1960, Herb Ross decided that the winter and early spring stoneflies of eastern North America would be excellent material for biogeographical and evolutionary studies, partly because they are easy to collect in large numbers. So he organized an informal 'Winter Stonefly Club' that eventually included about 150 members. He suggested that he would write up accounts of the most abundant genus, *Allocapnia*, and that I look after all the rest, this being about a 50:50 split, numerically. However, we both made contributions to all of the papers. Eventually these included revisions of the North American species of *Allocapnia* (127, 151),

Taeniopteryx (138) and *Zealeuctra* (144), and of the subfamily Brachypterinae throughout the world (165). We were surprised at the number of undescribed species that turned up, and the revisions required for old ones.

The Fraser River, especially its lower reaches, proved to be a rich source of aquatic insects. During the 1930s, a ferry crossed from Rosedale to Agassiz, and its docks were a good source of specimens. After the World War II a bridge replaced the ferry, and one of its concrete piers stood in shallow water where stonefly imagos would crawl out and remain in plain view. So this became the type of locality for four new species, and contributed large samples of several others.

There have been informal gatherings of stonefly workers every 3 or 4 years, starting with one in Switzerland in the 1950s. I have attended four of these; at Abisko in Swedish Lapland (on the border between forest and tundra); at Washington D.C.; at Tomahawk in northern Wisconsin; and most recently at Lausanne in Switzerland (Figure 14). At Washington I was asked to explain where my stonefly names had come from. Most of them were based on Russian, Spanish or Native American words that I was fairly sure would not duplicate anything already in use (175). Altogether, I have been the describer or co-describer of about 80 species that are now considered valid, and 43 subgenera that have been promoted to genera.

Evolution

During the 1920s an English mollusc expert named Robson wrote a book whose thesis was that species characters, as contrasted with those that distinguish genera and higher units of classification, are mostly not adaptive. This, he felt, was a defect of evolutionary theory. Robson's view frequently seems plausible if only morphological characters are considered. At a seminar I suggested that maybe the 'real' differences were in physiology or behaviour. Visible specific characters may often be either incidental or, especially among insects and birds, a device for reducing hybridization. Dymond thought I should publish a short note on this (3), though it seemed to me likely, then and now, that someone (probably Darwin) had had the same idea much earlier.

Figure 14. (a) Stan Szczytko of Wisconsin (left) and Bill Ricker (right) after the meeting of the stonefly researchers at Lausanne (1955). (b) Ian, Heidi, Bill Ricker, Teizi (left to right) after the meeting of the stonefly researchers at Lausanne (1955). (c) Kawaii (left) and Bill Ricker (right) after the meeting of the stonefly researchers at Lausanne (1955).

Since Darwin, I believe that there have been two and only two major breakthroughs in evolutionary theory, both of them during the 1920s. One was Alexander Oparin's extension of natural selection to account for the origin of life itself. The other was R.A. Fisher's brilliant reconciliation of Mendelian genetics with natural selection, two disciplines that had seemed to be completely opposed to each other. After initial astonishment and incredulity, both of these discoveries have succumbed to the Doctor Watson effect, and are now usually taken for granted. Modern workers are busy elaborating the details, which include the selection for body constituents that determine susceptibility or immunity to diseases and resistance to poisons, and selection for or against 'intelligence' (however that may be defined) in all kinds of vertebrate animals.

My later contributions to evolutionary lore have been incidental to other studies. Fisher's book prepared me to look for recently developed hereditary diversity among salmon stocks, and examples were close at hand. The various stocks of sockeye in the Fraser River watershed differ in morphology and in behaviour, yet all these differences must have evolved since the glaciers retreated, especially those directly related to survival (e.g., the direction in which the young fish have to swim in order to reach a suitable foraging region), or to success in reproduction (the fat content of the upstream migrating adults). I also postulated that the stock or stocks of kokanee (sockeye that live in fresh water) must have developed from anadromous sockeye in each lake separately (35, 107), and this has been confirmed by recent genetic analysis.

In part, the phylogeny of stoneflies presents a picture of decrease in size and in complexity of physical structure (fewer wing veins, loss of gills, reduction of genitalia) but, I surmise, a greater physiological efficiency (87). The exceptional group is the Perlidae, in which many species remain large, develop additional gills, and develop

a physiology that enables them to flourish in the tropics. A phylogenetic puzzle among the brachypterine. Taeniopterygidae was the occurrence, in several unrelated species, of a distinctive type of non-functional wings in the male. This is explicable on the view that a normally deleterious, hence rare, recessive mutation common to all brachypterines was selectively promoted to 'wild type' in species where mating occurs immediately after the females emerge from the water, so that it is advantageous that the males should not fly away (165). A similar situation occurs among stonefly populations that live at high elevations, where a nuptial flight often ends on a chilly snow bank from which there is no escape; or when an isolated population lives only in a short stretch of spring-fed stream that it is dangerous to leave. In such cases both sexes often have non-functional wings.

Physical, chemical and biological limnology

In the course of the Ontario trout work, determinations were made of the temperature and pH of the waters studied, and titrations of oxygen and carbonate content (5, 8). There was much diversity among the streams and small millponds. The big contrast was between the lime-rich waters of the Paleozoic sedimentary region of southern Ontario and the lime-poor brown waters of the Laurentian region. One thing that surprised me was that there was nearly always some supersaturation of oxygen in the upper levels of stratified waters, presumably a result of planktonic photosynthesis. This was so both in the millponds of the cuesta region and in the brown-water lakes of Algonquin Park.

At Cultus Lake we had 6 years of year-round observations of water temperature, chemistry, and plankton at intervals down through a column 40 m deep (20, 25), as well as collections of bottom invertebrates (89). The lake nearly always cooled below 4° C in winter, that is, past the temperature of maximum density; but complete circulation to the bottom usually continued down to 3.5° C, or in one year to 2.5° C, driven by cold northeast winds. This was followed by a period of partial circulation that usually lasted all winter. One year, however, the lake froze over completely and all circulation stopped. Similarly, in spring complete circulation continued past the point of maximum density to about 5.0° C, to be followed by partial circulation. At this time the thermocline began to develop near the bottom (not near the surface as the text books had it) and moved upward until fully developed, after which it was gradually pushed downward during the summer. However I failed to comment on this progression and only became aware of it as something new with E.M. Krokhin of Kamchatka referred to it – he found the same thing on Lake Dalnee near Petropavlovsk.

The plankton hauls were sorted to identify the organisms prevalent at each depth, and the crustacean species were compared with the numbers that were consumed by the young sockeye (21, 24, 25, 26).

Fish population dynamics as affected by harvesting

The technique of estimating the size of a population and its rate of harvest by marked members, begun in the 19th century by Petersen in Denmark, is now very widely used in biology and has an extensive literature. I employed it throughout the Indiana lake work, and was the first to develop an estimate of survival rate from recaptures (43, 54).

For the predator control work at Cultus Lake we need to know the absolute abundance of the important fish in the lake. A.N. Derzhavin had approached this problem by summing the catches taken from a sturgeon population over many years (152). My attempts to estimate populations by way of catch per unit of effort led to a discussion of the theory of that statistic (34, 47). It appeared that catch per unit of effort is of very limited usefulness when large differences in effort are being compared, so we repeated exactly the kinds and numbers of nets that were set in years at the start and end of the control program (38).

Rumours of a major work by F.I. Baranov in this field were intriguing enough that I got hold of a copy in Indiana and learned enough Russian to read it and eventually translate it. The rumours were correct: this paper, written in 1916, is still a basic reference in fish population analysis. It is centred around his 'catch equation':

$$C = NFA/Z$$

where C is catch in numbers, N is the number of fish present when fishing begins, A is the fraction of the N fish that die from all causes during the time of fishing, F is the instantaneous rate of fishing, and Z is the instantaneous rate of total mortality ($A = 1 - e^{-z}$).

My extensions of Baranov's work include the size limits paper (55) and the effects of environmental variability, stock mixtures and size-selective mortality on estimates of growth, mortality, production and yield (104, 125, 142, 152, 157, 167, 176).

The work of Petersen, Baranov, Derzhavin, Thompson & Bell, Graham, Beverton & Holt, Schaefer, Allen and other authors has been summarized in three handbooks, the so-called 'green books' (74, 105, 167). The last of these is available in French and English, and has been translated into Russian and published in the USSR. I also edited and contributed to the IBP Handbook (137), and wrote an historical chapter for John Gulland's compilation (174). The Green Books were, however, more than a review of published work. The first one, for example, included an examination of the effects of exploitation history on estimates of survival made from age composition. The later ones proposed a technique for estimating abundance and rate of harvest during recruitment years, whose mechanism is the same as for the 'cohort analysis' that became popular later. Another matter addressed in the books, and later given special treatment, is 'Lee's phenomenon'. When caused by mortality that is selective by the size of the fish within an age group, it can seriously distort estimates of both growth rate and mortality rate (142, 198).

The rate of growth of fishes is an important part of population dynamics. Two papers deal with growth rate of Indiana fishes (41, 42). Others compare growth rates with mortality rates for warm-water fishes (40, 43, 59, 99) and for salmon (120, 131, 173), and provide a synopsis of the mathematical aspects of growth technology (179).

Quantitative contribution of fishes to their ecosystem

Papers by V.S. Ivlev in the USSR and R.L. Lindeman in the USA drew attention to the concept of biological production (P) in aquatic habitats. Ivlev defined production as the total elaboration of organic matter by a stock during an interval of time; both by the organisms that survived to the end of the interval and by those that did not (135a). In the early 1940s the Ecological Society had a Symposium in this subject, and after much discussion we agreed that Ivlev's definition implied that production was equal to the instantaneous rate of growth of an organism of average size in an age-group (G) multiplied by the mean biomass of the age-groups that were present during the interval. For an exponential model of growth and mortality this can be written

$$P = G(B_1 - B_0)/(G - Z)$$

where B_0 and B_1 are the initial and final weight of the age-group and Z is its instantaneous rate of decrease in numbers (67, 68, 102, 177). Later this scenario was applied to the young sockeye in Cultus Lake, where there was quantitative information concerning their population. These fish were the 'key industry' in the pelagic region of the lake, and indeed in the lake as a whole (73).

As the work on the Indiana lakes progressed, it became clear that there was an abundance or overabundance of young bluegill sunfish being produced naturally (Figure 15). Hence there was no need for rearing young bluegills in hatchery ponds, maintained by the State and by numerous local clubs, or for a closed season on fishing up to

Figure 15. Bill Ricker, Indiana lake studies (1940s).

June 15 each year. Eventually the closed season was abolished, and my recommendation was that the hatchery ponds be used for fishing, or else to rear predacious species like largemouth bass, *Micropterus salmoides*, and pike, *Esox lucius*. This is not to say that there had been no human influence upon these fish populations. The wire traps that we used had formerly been a very popular method of catching a meal of sunfish, and we found that they caught different species with different efficiencies. Redear sunfish, *Lepomis microlophus*, pumpkinseeds, *Lepomis gibbosus*, and perch, *Perca flavescens*, were much more vulnerable than were bluegills. As a result, redears had become very scarce in most lakes, and in at least one lake where perch had been abundant we caught very few. Differences in vulnerability to cane-pole angling probably also made a contribution to these changes.

Stock and recruitment

During the 1930s, fish culturists and fishery managers seemed to be making one of two assumptions concerning this subject. One was that any increase in the eggs or young fish added to a population would produce a proportional increase in usable stock and hence harvest. So a vast network of fish hatcheries and rearing ponds, in Europe and North America, was pouring additional young fish into streams, lakes and even the oceans. The alternative assumption was that there would always be enough young fish from natural reproduction to saturate the biological productive capacity of a body of water, and all that was necessary was to harvest each stock most efficiently (Beverton & Holt 1957). It occurred to me that if these two situations existed, there must also be intermediate types, so in the early 1950s I made a series of trial calculations of possible relationships between the abundance of a fish stock and the average number of recruits that it might produce (90, 94). For one thing, this uncovered a mechanism that can maintain cycles of abundance in an animal population without environmental forcing. Cycles can arise whenever maximum recruitment occurs at less than the replacement level of reproduction, and will be permanent if the reproduction curve crosses the replacement line at an angle steeper than 45° (slope of –1). The kinds of cycles vary from a simple alternation in abundance to irregular or even 'chaotic' sequences. The author of a popular book on Chaos became aware of this work, but had me located in Australia! In populations consisting of two or more reproducing year-classes, the period of repetitive cycles proved to be twice the interval between a parental spawning and the average time of reproduction of its progeny (94, 146).

The best known aspect of this work is, however, the 'Ricker curve', which relates the number of recruits R to the abundance of their parents (P) at the same age:

$$\text{Log}_e(R/P) = a - bP$$

where e^a is the maximum value of R/P, which occurs when P is very small, and $1/b$ is the value of P that produces maximum absolute recruitment, which is e^{a-1}/b. This maximum can be at a stock size that is either smaller or larger than the replacement level (when $R = P$). Based on plausible ecological mechanisms in each case, the curve can have a wide range of possible shapes. It has been called 'the most parsimonious functional form', and serves as a standard against which observed relationships can be compared. Later, the effects of previous harvesting history were examined and found to be important (125, 157), and these effects were used to interpret the history of the Skeena sockeye populations (168). An alternative recruitment curve developed by Beverton & Holt (1957) is hyperbolic in shape and has an asymptotic maximum at the largest stock sizes. Jon Schnute (1985) has shown that both the Ricker and the Beverton–Holt curves are special cases of a 3-parameter expression that also describes intermediate and other types of curves. Any recruitment curve that includes the parameter a above, or a transformation of it, can be used to separate the 'compensatory' mortality rate, which increases as stock abundance increases, from the mortality rate that is independent of abundance (167, 191).

Cycles in the abundance of salmon

The off- and on-year cycles of abundance of some populations of salmon have always seemed most curious, although there are similar situations among insects. In some regions pink salmon are abundant and scarce in alternating years. On the Fraser River some sockeye stocks are abundant in only one year out of four, and this pattern is known to have existed as far back as the early 1800s. The interval corresponds to the dominant life span of the species in question. Up to about 1930 the common view was that physical catastrophes of some sort had knocked out the low years, and that they had been unable to recoup.

To Foerster and me at Cultus Lake it seemed obvious that some kind of interaction between successive generations must be involved (38). Among several possibilities for sockeye, an interaction involving lake predators seemed most likely (84), and during the 1950s Fred Ward of the Salmon Commission showed that this did exist at Shuswap Lake. Carl Walters and others at the University of British Columbia have discussed other possibilities, and have introduced the 'Ricker curve' into the picture, but the predation cycle is the only scheme that has, as yet, a demonstrated ecological basis (205). The nature of the interaction for pink salmon is not yet known, but cannibalism by returning adults upon outgoing young in the ocean is a possibility (119, 195).

Trends in size among salmon

During the 1950s I became aware that Chinook salmon had decreased in size since the 1920s, and wondered whether other species had similar trends. In 1972 Percy Wickett showed that coho had become smaller, and in 1973 Tom Bitton did the same for pinks. During 1975–1978 we computed the sizes of salmon caught in British Columbia and Alaska through 1975 (178, 180, 181, 182, 183, 185). During 1951–1975 chinook, coho, and pink salmon decreased seriously in size, Chinook in average age as well. Chum decreased a little in size but actually increased in age. Sockeye changed very little in mean size or age, except that the percentage of jacks (ocean age 2) increased greatly in some stocks. In populations that contained sockeye of both ocean ages 3 and 4, Dudley Foskett and Harry Godfrey had shown how gillnets selectively removed fish of intermediate size, leaving the smaller of the 3s and the larger of the 4s in excess as breeders. Thus over the years the ocean age 3s became smaller and the ocean age 4s became larger. Usually gillnets and trolls caught larger salmon than did seines in the same statistical area, but gillnetted chum were smaller than those seined because many of them were taken in nets of a mesh size more suitable for smaller species.

It seemed that all of these changed could be interpreted in terms of the selective characteristics of the fisheries, in combination with the different life-histories and growth patterns of the five species. It had been shown that both rate of growth and age of maturity in Pacific salmon were partly controlled by genetic factors that could be altered by selection (154), and the observed rate of selection was quantitatively sufficient to have caused the observed declines in size (178). Correlations of size with coastal temperatures, which decreased form 1960 to 1975, were weak, of variable sign, and non-significant statistically. However, there were two instances where unusual abundance of pink salmon in a particular area was associated with size (178).

After 1975 the northern Chinook salmon began to increase again in size, and by 1990 had regained the size they had in the 1950s, though they were probably not as large as they had been originally. Pink and coho stopped decreasing in size in the northern and central parts of British Columbia, and in some cases gained a little. In the south, however, the decrease in size of pinks and cohos continued and even accelerated (202). Coastal temperatures increased abruptly during 1976–1987, and in Alaska there was a great increase in abundance of sockeye and other species. A comprehensive explanation of these various changes may emerge from the concept of 'changes of regimes' throughout the Pacific Ocean (Beamish et al. 1999), together with knowledge of the, sometimes rather narrow, ranges of environmental conditions required by each species (Welch et al. 1998).

Predator control and sockeye salmon production

Starting in 1935, the Board approved an intensive netting program in Cultus Lake to see if sockeye production could be increased by reducing the numbers of their predators (Figure 16). We succeeded in reducing the adult pike minnow, *Ptychocheilus oregonensis*, population to about 20% of their original abundance without diminishing the trout appreciably. The sockeye responded with a

Figure 16. (a) Fingerling fence for counting yearling sockeye salmon, *O. nerka*, at Sweltzer Creek, Cultus Lake, British Columbia (spring 1933). (b) J. Lawrence McHugh counting yearling sockeye salmon, *O. nerka*, at fingerling fence on Sweltzer Creek, Cultus Lake, British Columbia (spring 1933). (c) Dolly Varden charr, *Salvelinus malma*, at Cultus Lake, British Columbia (1936 or 1939). (d) Gordon Thompson, (left) and Bill Baxter (right) with floating dock, net rack and boat used to reduce predaceous fishes in Cultus Lake, British Columbia (1936 or 1937).

3- or 4-fold increase in survival to the smolt stage, and subsequent increase in the return of the adults from the sea in 2 successive years (31, 38). These large spawnings, however, produced far more young than the lake could rear to normal smolt size, and the returns of adults from these undersized smolts were very poor (91a). There were indications also that the reduced number of adult pike minnow in the lake greatly increased the number of pike minnow recruits, possibly because the reduced adult stock consumed fewer of their own young. These recruits became potential consumers of sockeye when they reached age 4 (26). Suggestions for further experimentation included limitation of the number of sockeye spawners and fertilization of the lake, along with continued removal of the adult pike minnow, but the work was discontinued in 1943.

Restoring the upriver sockeye and pink salmon of the Fraser River

Except in the 'big' years of their 4-year cycle, the Fraser catches of sockeye decreased gradually from about 1900 onward, presumably because of a United States and Canadian fishery that was capable of capturing up to 95% of the incoming fish. Rock that was dumped or fell into the Fraser canyon in 1913 and 1914 obstructed salmon runs in those years, and unfortunately 1913 happened to be a 'big' year. The obstructions were largely removed or blasted downriver during the winter of 1914–1915. However, most of the sockeye salmon stocks did not quickly recover their former abundance. An exception was the late-running Lower Adams River population that reappeared in large numbers in 1926 because of the fortuitous coincidence of an unusual migration route and a light fishery. Two Chilko sockeye lines reappeared in numbers late in the 1920s, and most stocks began to increase when salmon traps were banned from Puget Sound waters after 1934. Yet at that point most upriver stocks were very much scarcer than they had been in the big years before 1913, and they were increasingly threatened by proposed hydroelectric developments.

During the 1920s and 1930s the Canadian Fisheries Research Board was restrained from attacking this problem, except for testing the fate of stocks transplanted from other watershed, because it was expected that at any moment an international organization would take over. The U.S. Senate finally ratified the treaty that established such a body in 1937, and the International Pacific Salmon Fisheries Commission began its work in 1938, with W.F. Thompson of Seattle as Director of Investigations. I spent the year 1938 on its staff, along with Earle Foerster, Jack Kask, Roy Jackson, Clinton Atkinson, Edward Whitesel and others. During the summer of 1938 the tagging party at Hell's Gate verified the pattern of difficult water levels there, and showed that some sockeye were permanently blocked and retreated to unsuitable tributaries downriver.

This and similar work was continued in subsequent years, with the result that the commission built fishways at Hell's Gate in 1945–1946, which eased the passage of salmon there. However, when a Bulletin concerning the work appeared in 1945, the data indicated that the big trouble had in fact been too much fishing (65). Yet all the emphasis in the Bulletin was on difficulty of passage; no mention was made of a need for greater escapements from the nets. As it turned out, in 1945, the first fishway year, there was no increase in the rate of recovery of the depleted runs; in fact there was an overall decrease in total upriver spawners. Heeding this warning, the Commission recommended shorter fishing seasons, and in 1946 a major increase in upriver spawners finally got under way (193).

The Hell's Gate fishways were of course a major achievement in design and construction. In addition to eliminating an annual loss of possibly 5–10% of the sockeye that reached them, they have permitted pink salmon to reestablish themselves. The upriver pink salmon had been almost exterminated in 1913, but began to reappear in numbers after the fishways were built. However, both upriver and downriver pinks suffered from overfishing that reached a climax in 1961. They gradually increased only when their rate of harvest was reduced (195).

Growth, natural mortality, and non-catch mortality of salmon at sea

During the post-war negotiations involving Canada, Japan, and the United States, an important consideration was whether harvesting salmon at sea produced smaller or a larger total catch than 'terminal' harvesting, when the fish had reached full growth. I undertook to examine this question for sockeye (120) and for pink and chum salmon (131). Later U.S. biologists produced several papers, and in 1974 I was asked to prepare a comprehensive review of this work (173). This included the effects of losses of fish from nets and longlines.

It turned out that ocean growth greatly exceeded natural mortality in the ocean, for all species. Hence greatest weight of a catch is obtained by harvesting near the end of the life span, in or near the river of origin, not during either the penultimate or the last year of ocean life while growth is still rapid. There were also important losses from salmon falling out of ocean gillnets as they are lifted, and from sharks or fur seals taking salmon caught by gillnets or longlines. In coastal waters there is some loss to seals and sea lions, except in seining, and there is a major 'shaker' loss of undersized salmon that are hooked and damaged by a troll and then die after release.

Pacific salmonids in the east

There is a tremendous contrast between the western and eastern parts of Canada in respect to their native salmonid fishes, as regards both variety and abundance. The Atlantic rivers produce only a few million kilos of salmon commercially, and provide river angling for a few sportsmen. Pacific rivers produce hundreds of millions of kilos commercially, and support major saltwater sport fisheries and river angling for coho and Chinook. Furthermore, the western side of the Pacific Ocean, which is climatically and oceanographically similar to the northwest Atlantic, historically produced just as many salmon as the eastern side.

In 1953, I spoke at Ottawa on the possibility and desirability of establishing Pacific-style stocks of salmon in the Atlantic; preferably pink salmon, whose life history does not overlap that of Atlantic salmon in fresh water. W.J.K. Harkness, who had moved to the Ontario Department of Lands and Forests, heard about this and decided to try putting pink and chum salmon into Hudson Bay. The bay proved to be too cold for salmon in winter; its native salmonids, brook trout and arctic charr, *Salvelinus alpinus*, are forced to retreat into river mouths in winter. However, some of the transplanted pink salmon escaped from the hatchery at Thunder Bay and got into Lake Superior. They are now established there and have moved into the other Great Lakes, much to everyone's surprise because there are no lake-dwelling pink populations in the west.

A little later the Fisheries Research Board tried to establish a run of pink salmon in a Newfoundland river. There were some reasonable returns from quite small transplants, but eventually the salmon disappeared. A more sustained effort, with initial protection from fishing, would likely be successful, as the Russians have succeeded against greater difficulties in the Murmansk region.

The first western salmonid to become common in eastern America was the rainbow trout. At the time of our work in the cuesta region of Ontario it had been established in only one river, the Pine, *oncorhynchus mykiss* but since then it and the brown trout, *Salmo trutta*, from Europe have superseded brook trout in all but the coldest upper reaches of most southern Ontario rivers. After lampreys had decimated the native lake trout in the upper lakes, the rainbows moved down into them and developed lake-living populations of large-sized fish, similar to the anadromous steelhead form in the Pacific. In the 1940s Michigan, and later other states and Ontario, imported large number of young coho and Chinook salmon into the Great Lakes. This was an enormous success, partly because the upper lakes had recently been denuded of salmonids by the sea lamprey, and were full of salmon foods like ciscoes, smelt and alewives.

Statistical methods

The plankton work at Cultus Lake required a statistical test of the accuracy of an individual count. Fiducial limits for the binomial distribution had recently been published, and with considerable diffidence I developed a table for the Poisson distribution by summing probabilities in Karl Pearson's table (16). This method was approved by E.S. Pearson, whom I saw in London in 1935, and the table has proved useful in a variety of contexts where the Poisson situation exists. That is when there are items or observations that have a low probability of occurrence, yet our information comes from those that in fact do occur. Another early problem was that of testing for the randomness of the distribution of plankters in the lake, and this was sorted out with the help of two letters from R.A. Fisher (15, 24).

More recently I felt a need to review the literature of linear regression, to see what statistic is appropriate when both variates of a pair are characterized by inherent variability, with or without measurement error as well. I also performed tests to see what happens when distributions are not bivariate normal (156, 186, 187). Both problems occur very widely in science, yet have not been sufficiently appreciated or satisfactorily handled. For example, in back-calculating fish lengths from marks on scales, etc., the use of ordinary regressions has led to consistent overestimation of first-year growth and underestimation of growth rate in later years (198). Twice I was asked to write on fishery applications of mathematics or statistics: once for the first volume of the *Biometrics Bulletin* (56), and more recently for the Fish Physiology series edited by W.S. Hoar and others (179).

Russian translations and dictionaries

When I returned to the west coast of Canada in 1950, little was known locally about Russian salmon stocks and Russian research on them. I started dictating translations of papers in this and other fields, sometimes at the request of colleagues. To keep track of these and other translations, we started listing them systematically in a 'Translation Series', which has been continued to the present. Nowadays most translations in Canada are done by or for a federal Bureau for Translations, which serves all Departments. I have written translation of about 110 Russian papers, or about 1000 pages total. After a year or two I started keeping a list of words and phrases not found in most dictionaries. Later, names of fishes and other aquatic animals and plants were added, making about 11000 definitions altogether. J.C. (Cam) Stevenson, who had become the Board's editor, was greatly interested in this and arranged for its publication (161). I was also asked to edit a translation of a dictionary by A.A. Klykov, a Russian fishery veteran, which contains a lot of regional and obsolescent names for fish and fishing gear (139).

Committees, Commissions and Conferences

At the first meeting of the Fisheries Research Board that I attended, the Chairman, G.B. Reed of Queen's University, suggested that, as Editor, I should make annual visits to the Board's research stations. These proved both interesting and very enjoyable, especially while the country still had comprehensive passenger rail services. They also permitted close acquaintance with the background and the personalities behind many of the papers I was to consider. And they proved most valuable when, later, I was asked to prepare or edit accounts of the Board's work for special occasions (115, 123, 126, 150, 160). The last of these items was an overview of the Board's work from its beginning to 1973, when it ceased to be a discrete operational unit of the Department of Fisheries.

Canada has become a member of a number of international commissions concerned with fisheries, and indeed was active in setting up some of them. Most of them meet annually, usually in the capital city of a member country. I have served as one of the advisors to the Canadian representative at a number of such meetings, held in Ottawa, Washington, Moscow or Tokyo, also once each in Copenhagen, Madrid and Guayaquil, Ecuador. These were concerned either with a special region (North Pacific, Northwest Atlantic) or with particular kinds of organisms (salmon, fur seals, whales, tropical tunas), and I often assisted with the preparation of documents (135, 164).

Farther afield, the FAO asked me to serve on a panel to review work done on the Peruvian anchoveta, *Engraulis ringens*, which had suddenly become the basis for a major fishery. The members included Luit Boerema and John Gulland of FAO, Garth Murphy, Gerald Paulik, Brian Rothschild, M.B. Schaefer and several others (the membership varied). We met in Lima, Peru four times and did some analyses of data, as well as recommending or commenting on research done by the local Instituto del Mar. I remember seeing my first hand-calculator there; Garth had just acquired it for about $400, and he predicted quite accurately that they would soon be available for $100 or less.

Preston Cloud (Paleontologist, Santa Barbara) invited me to be a member of the U.S. National Academy of Science's Committee on Resources and Man. This proved extremely interesting, particularly the four meetings with invited experts, and I contributed a chapter to the final report (145). Looking back 30 years later, the Committee's assessments and predictions have proved generally accurate, especially on the oil crunch (though it didn't forecast OPEC). It was overly optimistic on the future of nuclear power, predicting that breeder reactors would be on line by now and would soon provide abundant low-cost electricity.

From published observations and experiments I pointed out that the oceans are far from being an inexhaustible source of human food, as some had suggested. On the contrary, a limit to biological production of the kinds of fish taken by traditional fisheries was looming in the near future, so that the postwar expansion of such fisheries would soon slow down and come to a halt. However, overharvesting continued, so that several major stocks eventually became critically reduced – most recently the cod, *Gadus morhua*, of the northwest Atlantic.

Ecology

Back in the 1920s ecology was still a new branch of biology, and many academic institutions had yet to come to grips with it. Sometimes the word was still spelled oecology, from a Latin transcription of Greek oikos – a home. However the first 'o' soon disappeared, as it had earlier from the related term economics. In fact the word binomics had been proposed for much the same field as ecology, and by derivation this is perhaps the better term. But it is usage, not etymology, that determines whether a word lives or dies, and to date ecology has outdistanced all its competitors. The scientific newcomer of the 1920s began to appear publicly in the 1950s, and then rapidly became part of the everyday speech of millions. Soon it generated derivatives and compounds, like ecofriendly, ecotourism, ecofreaks, and many more.

Most of the work outlined above can be called ecological in a broad sense. Even that on taxonomy of stoneflies is a contribution to the cataloguing of living organisms, without which much confusion can arise. So it is as an ecologist that I would like to be remembered. If fish and fisheries are still flourishing 50 years from now, and salmon in particular, I like to feel that I will have made a small contribution to this happy result. But continual vigilance will be essential. Resource maintenance is under unremitting pressure from the standard business practice of estimating the present value of future supplies using a discount rate. Unfortunately this means that the fish to be caught 50 years from now, or even as little as 20 years from now are worth practically nothing today.

Personal bibliography of William E. Ricker

Since no complete listing of my publications has previously been published, I took this occasion to provide such a list. The numbered list below includes papers that had and may still have some importance in their respective fields. Prior to 1940, the Fisheries Research Board published brief popular articles concerning its work, the *Progress Reports*, to which all the scientists contributed from time to time. In Indiana I wrote, about twice a year, brief articles for the Department of Conservation's popular magazine *Outdoor Indiana*. After 1950, however, I was glad to leave this type of publicity to the Department of Fisheries' establishment in Ottawa, headed by 'Tommy' Turner and his successors. Especially in recent years, I have done some desultory writing on non-scientific matters, of which only No 194 and 208 are included below. So, in the end, this is still not a complete listing of all my publications.

(3) 1932. Physiological changes and the origin of species. Canadian Field-Naturalist 46: 30–31.

(4) 1932. Studies of speckled trout (*Salvelinus fontinalis*) in Ontario. University of Toronto Studies, Biological Series No. 36: 69–110. Publications of the Ontario Fisheries Research Laboratory, No. 44.

(5) 1932. Studies of trout producing lakes and ponds. University of Toronto Studies, Biological Series, No. 36: 113–167. Publications of the Ontario Fisheries Research Laboratory, No. 45.

(8) 1934. An ecological classification of certain Ontario streams. University of Toronto Studies, Biological Series, No. 37: 1–114. Publications of the Ontario Fisheries Research Laboratory, No. 49.

(11) 1935. [W.E. Ricker & E.B.S. Logier]. Notes on the occurrence of the ribbed toad (*Ascaphus truei* Stejneger) in Canada. Copeia, 1935: 46.

(12) 1935. [W.E. Ricker & A. Robertson]. Observations on the behaviour of adult sockeye salmon during the spawning migration. Canadian Field-Naturalist 49: 132–134.

(15) 1937. Statistical treatment of sampling processes useful in the enumeration of plankton organisms. Archiv für Hydrobiologie 31: 68–84.

(16) 1937. The concept of confidence or fiducial limits applied to the Poisson frequency distribution. Journal of the American Statistical Association 32: 349–356.

(20) 1937. Physical and chemical characteristics of Cultus Lake, British Columbia. Journal of the Biological Board of Canada 3: 363–402.

(21) 1937. The food and the food supply of sockeye salmon (*Oncorhynchus nerka* Walbaum) in Cultus Lake, British Columbia. Journal of the Biological Board of Canada 3: 450–468.

(24) 1938. On adequate quantitative sampling of the pelagic net plankton of a lake. Journal of the Biological Board of Canada 4: 19–32.

(25) 1938. Seasonal and annual variations in the quantity of pelagic net plankton, Cultus Lake, British Columbia. Journal of the Biological Board of Canada 4: 33–47.

(26) 1938. A comparison of the seasonal growth rates of young sockeye salmon and young squawfish in Cultus Lake. Biological Board of Canada, Progress Reports of the Pacific Coast Stations, No. 36: 3–5.

(27) 1938. [E.M. Walker & W.E. Ricker]. Notes on Odonata from the vicinity of Cultus Lake, B.C. Canadian Entomologist 70: 144–151.

(28) 1938. Notes on specimens of American Plecoptera in European collections. Transactions of the Royal Canadian Institute 22: 129–156.

(30) 1938. "Residual" and kokanee salmon in Cultus Lake. Journal of the Fisheries Research Board of Canada 4: 192–218.

(31) 1938. [R.E. Foerster & W.E. Ricker]. The effectiveness of predator control in decreasing the mortality of young sockeye salmon (*Oncorhynchus nerka* Walbaum). Verhandlungen der internationale Vereinigung für theoretische und angewandte Limnologie 8: 151–167.

(33) 1939. [W.E. Ricker & C.H.D. Clarke]. The birds of the vicinity of Lake Nipissing, Ontario. Contributions of the Royal Ontario Museum of Zoology, No. 16, 25 pp.

(34) 1940. On the relation of "catch per unit effort" to abundance and the rate of exploitation. Journal of the Fisheries Research Board of Canada 5: 43–70.

(35) 1940. On the origin of kokanee, a fresh-water type of sockeye salmon. Transactions of the Royal Society of Canada, Section V, 34: 121–135.

(36) 1941. The consumption of young sockeye salmon by predacious fish. Journal of the Fisheries Research Board of Canada 5: 293–313.

(38) 1942. [R.E. Foerster & W.E. Ricker]. The effect of reduction of predacious fish on survival of young sockeye salmon at Cultus Lake. Journal of the Fisheries Research Board of Canada 5: 315–336.

(40) 1942. [Karl F. Lagler & W.E. Ricker]. Biological fisheries investigations of Foots Pond, Gibson County, Indiana. Investigations of Indiana Lakes & Streams, 2: 47–72.

(41) 1942. [W.E. Ricker & Karl F. Lagler]. The growth of spiny-rayed fishes in Foots Pond, Indiana. Investigations of Indiana Lakes & Streams 2: 85–97.

(42) 1942. The rate of growth of bluegill sunfish in lakes of northern Indiana. Investigations of Indiana Lakes & Streams 2: 161–214.

(43) 1942. Creel census, population estimates and rate of exploitation of game fish in Shoe Lake, Indiana. Investigations of Indiana Lakes & Streams 2: 215–243

(45) 1943. Stoneflies of southwestern British Columbia. Indiana University Publications, Science Series, No. 12, 145 pp.

(47) 1944. Further notes on fishing mortality and effort. Copeia 1944: 23–44.

(49) 1944. Some Plecoptera from the far north. Canadian Entomologist 76: 174–185.

(53) 1945. [W.E. Ricker & Daniel Merriman]. On the methods of measuring fish. Copeia 1945: 185–191.

(54) 1945. Natural mortality among Indiana bluegill sunfish. Ecology 26: 111–121.

(55) 1945. A method of estimating minimum size limits for obtaining maximum yield. Copeia 1945: 84–94.

(56) 1945. Some applications of statistical methods to fishery problems. Biometrics Bulletin 1: 73–79.

(57) 1945. A first list of Indiana stoneflies. Proceedings of the Indiana Academy of Science 54: 225–230.

(59) 1945. Abundance, exploitation and mortality of the fishes in two lakes. Investigations of Indiana Lakes & Streams 2: 345–448.

(65) 1947. Hell's Gate and the sockeye. Journal of Wildlife Management 11: 10–20.

(66) 1946. Some prairie stoneflies (Plecoptera). Transactions of the Royal Canadian Institute 26: 3–8.

(67) 1946. Production and utilization of fish populations. Ecological Monographs 16: 373–391.

(68) 1946. [G.L. Clarke, W.T. Edmondson & W.E. Ricker]. Mathematical formulation of biological productivity. Ecological Monographs 16: 336–337.

(72) 1948. Stoneflies of the maritime provinces and Newfoundland. Transactions of the Royal Canadian Institute 26: 404–414.
(73) 1948. [W.E. Ricker & R.E. Foerster]. Computation of fish production. Bulletin of the Bingham Oceanographic Collection 11: 173–211.
(74) 1948. Methods of estimating vital statistics of fish populations. Indiana University Studies, Science Series, No. 15, 101 pp.
(82) 1949. The North American species of Paragnetina (Plecoptera, Perlidae). Annals of the Entomological Society of America 42: 279–288.
(83) 1949. Mortality rates in some little-exploited populations of freshwater fishes. Transactions of the American Fisheries Society for 1947, 77: 114–128.
(84) 1950. Cycle dominance among the Fraser sockeye. Ecology 31: 6–26.
(87) 1950. Some evolutionary trends in Plecoptera. Proceedings of the Indiana Academy of Science 59: 197–209.
(88) 1952. Systematic studies in Plecoptera. Indiana University Publications, Science Series, No. 18, 200 pp.
(89) 1952. The benthos of Cultus Lake. Journal of the Fisheries Research Board of Canada 9: 204–212.
(90) 1952. Numerical relations between abundance of predators and survival of prey. Canadian Fish Culturist 13: 5–9.
(91) 1953. [R.E. Foerster & W.E. Ricker]. The coho salmon of Cultus Lake and Sweltzer Creek. Journal of the Fisheries Research Board of Canada 10: 293–319.
(91a) 1953. [R.E. Foerster & W.E. Ricker]. Predator control and sockeye salmon production. Fisheries Research Board of Canada Manuscript Report No. 478, 18 pages.
(94) 1954. Stock and recruitment. Journal of the Fisheries Research Board of Canada 11: 559–623. [Russian translation, ìZapas I popolnenieî, in Materialy Mezhdunarodnoi Konferentsii po Okhrane Zapasov Ryb 1: 51–93. Moscow, 1957].
(98) 1955. [W.E. Ricker & W.W. Judd]. A collection of stoneflies (Plecoptera) from the Northwest Territories, Canada. Entomological News 66: 255–256.
(99) 1955. Fish and fishing in Spear Lake, Indiana. Investigations of Indiana Lakes & Streams 4: 117–161.
(102) 1958. Production, reproduction and yield. Verhandlungen der internationale Vereinigung für theoretische und angewandte Limnologie 13: 84–100.
(104) 1958. Maximum sustained yields from fluctuating environments and mixed stocks. Journal of the Fisheries Board of Canada 15: 991–1006.
(105) 1958. Handbook of computations for biological statistics of fish populations. Fisheries Research Board of Canada, Bulletin No. 119, 300 pp.
(106) 1959. The species of *Isocapnia* Banks (Insecta, Plecoptera, Nemouridae). Canadian Journal of Zoology 37: 639–653.
(107) 1959. Additional observations concerning residual sockeye and kokanee (*Oncorhynchus nerka*). Journal of the Fisheries Research Board of Canada 16: 897–902.
(108) 1959. Plecoptera. pp. 941–957. In: H.B. Ward & G.C. Whipple (editors) Fresh – Water Biology, 2nd ed. John Wiley, New York.
(209) 1960. A population of dwarf coast range sculpins (*Cottus aleuticus*). Journal of the Fisheries Research Board of Canada 17: 929–932.
(115) 1961. Productive capacity of Canadian fisheries: an outline. Background papers, Resources for Tomorrow Conference 2: 775–783. Ottawa. [Condensed version in Trade News 14: 5–7. Canada Department of Fisheries].
(117) 1962. [W.E. Ricker & Ferris Neave]. Nesting colony of mew gulls of Kennedy Lake, Vancouver Island. Report of the British Columbia Provincial Museum of Natural History & Anthropology for 1961, pp. 20–21.
(119) 1962. Regulation of the abundance of pink salmon populations. University of British Columbia, H.R. MacMillan Lectures in Fisheries, Symposium on Pink Salmon, 1960, pp. 155–201.
(120) 1962. Comparison of ocean growth and mortality of sockeye salmon during their

last two years. Journal of the Fisheries Research Board of Canada 19: 531–560.

(123) 1962. [Editor]. Productive capacity of Canadian fisheries. Fisheries Research Board of Canada, Nanaimo Biological Station Circular No. 64, 79 pp.

(125) 1963. Big effects from small causes: Two examples from fish population dynamics. Journal of the Fisheries Research Board of Canada 20: 257–264.

(126) 1964. [P.A. Larkin & W.E. Ricker, editors]. Canada's Pacific Marine Fisheries: past performance and future prospects. pp. 194–268. Inventory of the Natural Resources of British Columbia, Victoria, B.C. [Also issued separately in 1963 for the 15th British Columbia Natural Resources Conference.]

(127) 1964. [H.H. Ross & W.E. Ricker]. New species of winter stoneflies, genus *Allocapnia* (Plecoptera, Capniidae). Transactions of the Illinois Academy of Science 57: 88–93.

(131) 1964. Ocean growth and mortality of pink and chum salmon. Journal of the Fisheries Research Board of Canada 21: 905–931.

(132) 1964. Distribution of Canadian stoneflies. Gewasser und Abwasser 34/35: 50–71.

(133) 1965. New records and descriptions of Plecoptera (Insecta). Journal of the Fisheries Research Board of Canada 22: 475–501.

(135) 1966. Salmon of the North Pacific Ocean-Part III. A review of the life history of North Pacific salmon. 4. Sockeye salmon in British Columbia. , pp. 59–70. Bulletin of the International North Pacific Fisheries Commission, No. 18.

(135a) 1966. (Footnotes to a translation of V. S. Ivlev. The biological productivity of waters.) Journal of the Fisheries Research Board of Canada 23: 1727–1759.

(137) 1968. Introduction (Chapter 1), Appendices, and Index. pp. 1–6, 255–269 and 307–313 In: W.E. Ricker (editor) Methods for assessment of fish production in fresh waters. IBP Handbook No. 3. Blackwell Scientific Publications, Oxford.

(138) 1968. [W.E. Ricker & H.H. Ross]. North American species of *Taeniopteryx* (Plecoptera, Insecta). Journal of the Fisheries Research Board of Canada 25: 1423–1439.

(139) 1968. [Editor]. A short dictionary of fishery terms (Russian-English). Compiled by A. A. Klykov; translated and published by U.S. Department of Commerce, Clearinghouse for Federal Scientific & Technical Information, Joint Publications Research Service. 67 pp.

(140) 1968. [W.E. Ricker, R. Malouin, P. Harper & H.H. Ross]. Distribution of Quebec stoneflies (Plecoptera). Le Naturaliste Canadien 95: 1085–1123.

(142) 1969. Effects of size-selective mortality and sampling bias on estimates of growth, mortality, production, and yield. Journal of the Fisheries Research Board of Canada 26: 479–541.

(144) 1969. [W.E. Ricker & H.H. Ross]. The genus *Zealeuctra* and its position in the family Leuctridae (Plecoptera, Insecta). Canadian Journal of Zoology 47: 1113–1127.

(145) 1969. Food from the sea. pp. 87–108. In: P. Cloud (editor). Resources and man, a study and recommendations. Report of the Committee on Resources and Man, U.S. National Academy of Sciences, W.H. Freeman & Co., San Francisco.

(146) 1969. [Arthur Tautz, P.A. Larkin & W.E. Ricker]. Some effects of simulated long-term environmental fluctuations on maximum sustained yield. Journal of the Fisheries Research Board of Canada 26: 2715–2726.

(150) 1970. [P.A. Larkin, D. Pimlott & W.E. Ricker]. Vertebrate ecology: a review of present status. Report No. 5, 40 pages, in Panel Reports of the Study of Basic Biology in Canada by the Biological Council of Canada and the Canadian Federation of Biological Sciences. Biological Council of Canada, Ottawa.

(151) 1971. [H.H. Ross & W.E. Ricker]. The classification, evolution and dispersal of the winter stonefly genus *Allocapnia*. Illinois Biological Monographs No. 45, 166 pp.

(152) 1971. Derzhavins's biostatistical method of population analysis. Journal of the Fisheries Research Board of Canada 28: 1666–1672.

(154) 1972. Hereditary and environmental factors affecting certain salmonid populations. p. 27–160. In: The Stock Concept in Pacific Salmon. H.R. MacMillan Lectures in Fisheries, University of British Columbia, Vancouver.
(155) 1972. [A.R. Gaufin, W.E. Ricker, M. Miner, P. Milam & R.A. Hays]. The stoneflies (Plecoptera) of Montana. Transactions of the American Entomological Society 98:1–161.
(156) 1973. Linear regressions in fishery research. Journal of the Fisheries Research Board of Canada 30: 409–434.
(157) 1973. Two mechanisms that make it impossible to maintain peak-period yields from stocks of Pacific salmon and other fishes. Journal of the Fisheries Research Board of Canada. 30: 1275–1286.
(161) 1973. Russian-English dictionary for students of fisheries and aquatic biology. Bulletin of the Fisheries Research Board of Canada 183: 1–428.
(164) 1974. [W.E. Ricker & J.I. Manzer]. Recent information on salmon stocks in British Columbia. Bulletin of the International North Pacific Fisheries Commission 29:1–36.
(165) 1975. [W.E. Ricker & H.H. Ross]. Synopsis of the Brachypterinae (Insects, Plecoptera, Taeniopterygidae). Canadian Journal of Zoology 53: 132–153.
(167) 1975. Computation and interpretation of biological statistics of fish populations. Bulletin of the Fisheries Research Board of Canada 191: 1–382.
(168) 1975. [W.E. Ricker & H.D. Smith]. A revised interpretation of the history of the Skeena River sockeye salmon. Journal of the Fisheries Research Board of Canada 32: 1369–1381.
(169) 1975. The Fisheries Research Board of Canada: seventy-five years of achievements. Journal of the Fisheries Research Board of Canada, 32: 1369–1381.
(170) 1974. [A.R. Gaufin & W.E. Ricker]. Additions and corrections to a list of Montana stoneflies. Entomological News 85: 285–288.
(171) 1975. W.E. Ricker & G.G. Scudder. An annotated checklist of the Plecoptera (Insecta) of British Columbia. Syesis 8: 285–288.
(173) 1976. Review of the rate of growth and mortality of Pacific salmon in salt water, and noncatch mortality caused by fishing. Journal of the Fisheries Research Board of Canada 33: 1483–1524.
(174) 1977. The historical development. pp. 1–26 In: J.A. Gulland (editor).Fish population dynamics, John Wiley & Sons, London.
(175) 1976. Origin of stonefly names proposed by Ricker and collaborators. Perla 2: 5–19. Brigham Young University, Provo, Utah. Reprinted with additions and emendations, 1992.
(176) 1978. [S.J. Westrheim & W.E. Ricker]. Bias in using an age-length key to estimate age-frequency distributions. Journal of the Fisheries Research Board of Canada 35: 184–189.
(177) 1978. On computing production. Limnology & Oceanography 23: 379–380.
(178) 1978. [W.E. Ricker, H.T. Bilton & K.V. Aro]. Causes of the decrease in size of pink salmon (*Oncorhynchus gorbuscha*). Canadian Fisheries & Marine Service Technical Report No. 820, 93 pp.
(179) 1979. Growth rates and models. pp. 677–743. In: W.S. Hoar & D.J. Randall (editors) Fish Physiology, volume 8, Academic Press, New York.
(180) 1980. Changes in the age and size of chum salmon (*Oncorhynchus keta*). Canadian Technical Report of Fisheries & Aquatic Sciences, No.930, 99 pages.
(181) 1980. Causes of the decrease in age and size of Chinook salmon (*Oncorhynchus tshawytscha*). Canadian Technical Report of Fisheries & Aquatic Sciences, No. 944, 25 pages.
(182) 1980. [W.E. Ricker & W.P. Wickett]. Causes of the decrease in size of coho salmon (*Oncorhynchus kisutch*). Canadian Technical Report of Fisheries & Aquatic Sciences, No. 971, 63 pages.
(183) 1981. Changes in the average size and average age of Pacific salmon. Canadian

Journal of Fisheries & Aquatic Sciences 38: 1636 – 1656.
(185) 1982. Size and age of British Columbia sockeye salmon (*Oncorhynchus nerka*) in relation to environmental factors and the fishery. Canadian Technical Report of Fisheries & Aquatic Sciences, No. 1115, 117 pages.
(186) 1982. Linear regressions for naturally variable data. Biometrics 38: 859 – 860.
(187) 1984. Computation and uses of central trend lines. Canadian Journal of Zoology 62: 1897 – 1905.
(191) 1985. Varieties and properties of three recruitment curves. La Mer [Tokyo] 23: 153 – 157.
(193) 1987. Effects of the fishery and of obstacles to migration on the abundance of Fraser River sockeye salmon (*Oncorhynchus nerka*). Canadian Technical Report of Fisheries & Aquatic Sciences No. 1522, 75 pages.
(194) 1988. The absent corpse mystery solved; or, Professor Moriarty in the Yukon. Canadian Holmes 12: 3 – 11.
(195) 1989. History and present state of the odd-year pink salmon runs of the Fraser River region. Canadian Technical Report of Fisheries & Aquatic Sciences No. 1702, 37 p.
(198) 1992. Back-calculation of fish lengths based on proportionality between scale and length increments. Canadian Journal of Fisheries & Aquatic Sciences 49: 1018 – 1026.
(201) 1994. (P.P. Harper & W.E. Ricker) Distribution of Ontario stoneflies. Proceedings of the Entomological Society of Ontario 125: 43 – 66.
(202) 1995. Trends in the average size of Pacific salmon in Canadian catches. Canadian Special Publication in Fisheries & Aquatic Sciences 121: 593 – 602.
(205) 1997. Cycles of abundance among Fraser River sockeye salmon (*Oncorhynchus nerka*). Canadian Journal of Fisheries & Aquatic Sciences 54: 950 – 968.
(206) 1997. (Kenneth W. Stewart & William E. Ricker). Stoneflies (Plecoptera) of the Yukon. pp. 201 – 222. *In*: H.W. Danks & J.A. Downes (editors). Insects of the Yukon. Biological Survey of Canada, Monograph No. 2, 1034 pages.
(208) 1999. (William E. Ricker & Jon T. Schnute). A westward trajectory extension for the celestial fireballs seen in 1913. Canadian Field-Naturalist 113: 693 – 697.

References cited

Beamish, R.J., D.J. Noakes, G.A. McFarlane, L. Klyashtorin, V.V. Ivanov & V. Kurashov. 1999. The regime concept and natural trends in the production of Pacific salmon. Canadian Journal of Fisheries & Aquatic Sciences 56: 516–526.

Beverton R.J.H. & S.J. Holt. 1957. On the dynamics of exploited fish populations. U.K. Ministry Agriculture, Food & Fisheries, Fisheries Investigations (Ser. 2), 19. 533 pp.

Dymond J.R., J.L. Hart & A.L. Pritchard. 1929. The fishes of the Canadian waters of Lake Ontario. University of Toronto Studies 33: 1 – 35. Publication of the Ontario Fisheries Research Laboratory No. 37.

Fry F.E.J. 1937. The summer migration of the cisco, Leucichthys artedi (Le Sueur), in Lake Nipissing, Ontario. University of Toronto Studies in Biology 44: 191. Publication of the Ontario Fisheries Research Laboratory No. 55.

Schnute, J.T. 1985. A general theory for analysis of catch and effort data. Canadian Journal of Fisheries & Aquatic Sciences 42: 414–429.

Welch, D.W., Y. Ishida & K. Nagasawa. 1998. Thermal limits and ocean migrations of sockeye salmon (Oncorhynchus nerka): long-term consequences of global warming. Canadian Journal of Fisheries & Aquatic Sciences 55: 937–948.

Environmental Biology of Fishes (2006) 75:39–72
DOI 10.1007/s10641-005-2442-y

Flora was his interest and prime course of study: a botanical career for W.E. Ricker disappears

Karl E. Ricker
3253 Archibald Way, V0N 1B3, Whistler, BC, Canada (e-mail: accutypeservices@shaw.ca–Attention Ricker)

Received 23 April 2004 Accepted 18 May 2004

Key words: plants, botany, ecology

Synopsis

Bill Ricker's career went through many twists in his academic years. He had taken botany in his senior matriculation year at high school and he had collected over 100 species of flora before commencement of university life. At the conclusion of his first university year, he set out over the summer to collect a much larger sample of species, primarily from the Great Lakes-St. Lawrence ecoregion, to fulfil a requirement for a second year botany course (spermatophytes). He identified about 390 species, and some 254 were collected and pooled with those from previous years to make a final submission of 354 spermatophyte species. Field plant identification continued in each academic year thereafter, in concert with collections and identifications of aquatic invertebrates in his summer projects while under the employment of the Ontario Fisheries Research Laboratory. At the conclusion of his undergraduate years, Bill had taken more courses in botany than in zoology, and it was the summer employment that had really prepared him for postgraduate work in fisheries biology, which was ecologically oriented. When Bill left Ontario in the autumn of 1931 he had identified over 600 species of plants, excluding lower cryptogams, but including many aquatic species of higher plants. In western North America Bill's botanical career began at Cultus Lake in 1931. He again studied all aspects of the basin while employed with the federal government, and from the work he assembled a Ph.D. thesis. At the time of thesis completion he had identified over 300 species of flora, including alpine plants at timberline, 1500 – 1800 m above lake level, and planktonic algae in its water column. In 1939, after more field fisheries work in the Fraser River basin of British Columbia, Bill accepted a position with the biological staff at Indiana University. In this period which concluded in 1950 he identified another 50 – 110 species of flora, all in the Carolinian ecoregion, and hitherto not seen by him. Considering all floral classes, Bill's eastern North American repertoire had by then added up to 791 species, representative of more than 112 families of plants. Returning west for the remainder of his life, new identifications elsewhere added to his Cultus Lake list which slowly added up to about 1000 species for the west coastal region of North America. Flora was also identified elsewhere in the mid-continental region of North America, in Eurasia where the Abisko region of Lappland was a highlight, and in South America and New Zealand. Records of his botanical prowess, were kept primarily in his diaries, which began in 1923 and were maintained consistently to the end of 1934, and thereafter intermittently to 1949. The diaries reveal that his career as a budding botanist was subtly hijacked by a wily Professor W.H.K. Harkness in the rival Biology Department who outmanoeuvred Drs. R.B. Thompson and R.A. Sifton in the Botany Department. The former always managed to employ Bill in summer and keep him occupied in the department's labs during the autumn and winter and spring, tying up any free time when the botanist had approached him on lab work. Certainly, the botany courses taken and which he excelled at were more appropriate for his aquatic ecological pursuits. Salesmanship won the day for the zoologists, but Bill was a life-long botanist regardless of whatever else he studied or managed throughout his professional career. The last days of his life had a botanical conclusion.

Introduction

Bill Ricker's life was one of continuous curiosity with an insatiable appetite to study all facets of nature, be it physical, biological, chemical or even metaphysical. The quest for knowledge began in his youthful years, searching the skies to study the positions and movements of celestial bodies. His father, Harry Edwin Ricker, was the 'science master' at the collegiate (normal school) in North Bay, Ontario and he taught his three daughters and one son, Bill the second oldest, to observe and appreciate their surroundings.

As a boy, however, railways fascinated him and to become a steam locomotive engineer was looked upon as an exciting career. By the time he reached high school, the allure of driving a locomotive was forgotten, and by age 13 he had memorized the star charts and could point out many a constellation in the sky no matter how obtuse it was to follow its outlined projectory from star to star.

In those early years Bill was also learning to identify the local avian fauna, thanks to some outside help arranged by his father and older sister, Evelyn. Bill was also instilled to keep written records of what he saw and what other unusual events took place during the day. Initially the record was of a list of birds seen on each outing, without details of where or when. In 1923, at age 14, the observations were written in diary format, focusing on birds, their behaviour and surroundings. Usually the diary entries fell on weekend days when Bill and pal, Charlie Wagar, were roaming the countryside. This early diary record provides the first hints of his awareness of the flora through which they rambled.

High school summers saw the Ricker family go south to the grandparents' farm at Plattsville, Oxford County, near Woodstock, Ontario. The transition from the Boreal Forest, lying on the granitic and acidic terrain about North Bay, to the Great Lakes-St. Lawrence floral region on the flatter farmlands, underlain by Paleozoic carbonates, was like night to day. Major differences in the biota of the two regions were shown by the lists of birds, plants and other organisms that Bill methodically compiled. By 1924, and 16 years old at summer's end, he was actively identifying the flora in each region, helped by older sister, Evelyn, who also had similar interests. The fourth year at high school, that is Form IV in those days, concluded the diary entries which focused mainly on birds. In the spring of that school year (1924/1925), he and Evelyn began to collect flowers for exacting identification as well as for long-term archival purposes. By this time she had her prized copy of Gray's New Manual of Botany (Robinson & Fernald 1908) which cost a staggering $3.25; she was now at the University of Toronto and had enrolled in the introductory botany course. Collections of flora were made at North Bay and at Plattsville, which continued into 1926 when Bill completed Form V (grade 13 of later years, senior matriculation, or first year university), taking Botany as one course in his final year at high school. By this time the lists of flora for the Plattsville region began to show a Carolinian floral element, an area which Scoggan (1978) has recognized as a local transition zone between the two ecoregions.

Academically Bill was a genius, although he never boasted of this to anyone including his sons. In fact, it was not until he passed away in 2001, age 93, when the family discovered that he was the gold medalist of his class, every year, at high school. While sorting out his estate the medals were found; report cards were retrieved from his files, and a stunning document was uncovered: 'University of Toronto – Award of Matriculation Scholarships' with an attached page labelled 'Matriculation Scholarship Examination – June, 1926'. The title page indicated that Bill received 'The First Edward Blake Scholarship' in science. In those days enrolment at the university required the mandatory writing of provincial exams at the conclusion of high school. Bill's expanded diary of 1926, for the months of May and June, did not place great confidence at obtaining first-class marks in those exams, and, in fact he and pal, Charlie, has spent the weekends during examination time rambling through the environs adjacent to North Bay, rather than focusing on concerted study. The second page of the University's document dispelled that diary myth with the following results for Bill:

- Science #1 in the province, and First Class (i.e. 'A')
- Physics #1 in the province, and First Class

- Chemistry #2 in the province, and First Class
- Biology #3 in the province, and First Class
- Algebra/Geometry #4 in the province, and Second Class (a tough exam!)
- History #5 in the province, and First Class
- German #5 in the province, and First Class
- Latin #33 in the province, and First Class
- Trigonometry #41 in the province, and First Class

Obviously, it was a 'slam-dunk' scholarship award!

The exam results and Bill's broad interest in science presented a conundrum. He already had well-grounded starts in ornithology, botany and astronomy, and he was obviously very good at physics, chemistry and languages and could hold his own in mathematics. Which pursuit would he attack at the University of Toronto? What would be the courses to be his life-long profession? Well, he did not know when he enrolled at the University (Honours Science) in the autumn of 1926 at age 18. For the first term, which ended in mid-January 1927, he signed up for chemistry (2 labs each week), two physics courses (Heat, Mechanical), Calculus, Geology, Scientific German (after toying with the idea of taking Greek), compulsory physical education (swimmer's life saving), and botany. Zoology was not even in the course of study! In the second term, however, general biology (zoology, by the appearance of the lab book) appears to have replaced the calculus, or was added to his curriculum; the lab book for the course revealed a very diluted exposé of general zoology – hardly a challenge to one mired in difficult physical science courses!

For the second year at university the same line of courses were taken; botany (Spermatophytes), zoology (Invertebrates), physics (Light, Electricity, Sound), Physical and Organic chemistry(!), Invertebrate Paleontology (with Historical Geology), and Scientific German. Mathematics appears to have been left out, but at that time it looked as if he was still heading to a career in either the physical sciences, or in botany. He had spent the entire previous summer collecting specimens for the botany course requirements of second year.

By third year a shift was evident; he enrolled in Psychology, dropped the physical sciences in favour of Biochemistry, and was a star student in Plant Physiology. But he was also making detailed drawings in Lower Cryptogamic Botany (2nd term), Comparative Anatomy and Embryology. A career in botany still appeared to be the course of direction; fisheries biology and the mathematics associated with it was not in the academic focus – yet.

The final undergraduate year saw three more botany courses taken: Upper Cryptogamic Botany, Botany Seminar, and Plant Ecology. There was also the nuisance course, 'History of Biology', but sliding into the picture was Histology, Genetics, Hydrobiology and Systematic Zoology (2 labs each week). Furthermore, he was a lab demonstrator in Comparative Anatomy (3 labs each week) and his own lab books in the course show why he excelled at this. On Saturday mornings he was also a lab demonstrator for 'oh joy': botany for engineering students!

At graduation the report card was again outstanding: Biology – number one and first class; History of Biology was only a 'B': yet number one in that class. The graduation also netted two awards: the gold medal in science, and a bronze from the British Association for the Advancement in Science. Again, it appeared that the net result of all academic endeavours was towards botany, although Dr. E.M. Walker was also prodding him to become an academic entomologist.

Behind the scenes throughout all academic years, however, was Professor W.H.K. Harkness who kept Bill employed in all summers with a jointly funded and operated Ontario Fisheries Research Laboratory that he was in charge of. For good measure he also provided some winter lab work to Bill, analyzing the food content of trout stomachs. Realizing that Bill's forté each summer was taxonomic astuteness in botany, including the more difficult aquatic taxons, and invertebrates, he directed Bill's strengths toward ecological appraisals of trout-bearing stream and private fish pond investigations, all located to the north-west of Toronto. Hence there was no problem at graduation to steer him onto a master's thesis project at Mad River, which occupied the summer of 1930 as well as some field trips during the academic year of 1930 – 1931.

From Bill's diaries for this first graduate study period, which do have exacting details of the field work, it is not clear what courses he simultaneously undertook at university. However, Vertebrate Paleontology was one, as shown by the laboratory book, which also outlines the geology of southern Ontario in some detail. Unfortunately, many laboratory books, course notes and text books for much of Bill's years at the university had been misplaced when his office was moved at the Pacific Biological Station in 1969, while he was away in the USSR. Those books that did not lose their way, however, show remarkable prowess in the academic laboratories, providing unbelievable detail. In fact, some jokester re-labelled his 'Comparative Anatomy Laboratory Drawings', III Honours biology, University of Toronto to 'Wrecker's Comparative Anatomy (Made Difficult) for Students, 1st Edition, All writes reserved'. To say the least, his lab books for both courses in cryptogamic botany are of similar standards.

At the conclusion of the Mad River work the career path for Bill was not settled, and in the summer of 1931 he did odd biological assignments for Dr. Harkness as far north as Kapuskasing, Ontario, and took a couple of weeks off, hitchhiking to Pt. Pelee and noting the flora wherever he went. The good professor finally made connections with the Biological Board of Canada which saw a job offer for Bill, and off to Chilliwack, British Columbia by train he went in mid-October to begin a study of Cultus Lake and its salmonid populations. Bill revelled in this new mountain basin environment, again identifying everything he saw and keeping up his meticulous diaries while doing so. Arrival in British Columbia was akin to heaving a weight off one's back and starting anew, but using the principles of his well indoctrined scientific background and outdoorsmanship he overcame the new adversities or unknowns. It was the Cultus Lake project (1931–1935) that shaped Bill's career, provided him with a wife (and family), and propelled the impetus to keep stepping out into the scientific unknowns.

In late 1935 Bill returned to the University of Toronto for his final academic stand, to quickly tidy up the thesis and to undertake the Ph.D. orals, followed by disembarkation on a 6 month sabbatical to Europe. From country to country he wandered to meet potential colleagues in his fields of interest, and they were invaluable in his scientific quests for years thereafter. The diary rigidly identified them all as well as recorded his other observations throughout the day. Returning to Cultus Lake, after the overseas networking was completed, the diaries report a daily operational life to the end of 1937, when changes in salmon politics took wind. Now employed with the newly-founded International Pacific Salmon Commission in 1938 there was no longer time to write diaries. The field work was exhaustively long each day and the diary or field book was filled with sketches of spawning beds, fish counts and operational logistic notes.

Unfortunately, the era of writing in the hard-backed diaries was coming to a close when he moved to Indiana in 1939, although there were intermittent, short attempts to revive it (e.g. 1948–1949). Trip reports were prepared in their place, if the venture was sustained or interesting. About 30–40 such reports were written over the time span of 1946–1995, many of which mention natural history observations as well as the day's events or direction of travel.

For this review I have used those diaries and trip reports extensively to delve into Bill's career as a botanist (in absentia), noting the highlights, the biological observations, and other fascinations in them. The intent of this review is to show the intensity and whereabouts of his botanical endeavours and summarize the numbers of taxa identified in some areas of his special interest. In his university years he prepared lists of plants identified and of those collected, but in many cases the diaries provide additional observations to embellish the species count, as recompiled by the author.

After the conclusion of Ph.D. work the listing of botanical identifications was halted, as were any further identifications of pesky grasses, sedges and rushes. More often the identification of a plant was ticked off in a floral identification manual with a short note on the place and date. For eastern North America he used his sister's copy of Gray's New Botany Manual, printed at the ripe old age of 1908; for western Canada and USA it was Henry's (1915) manual, but eventually replaced by Peterson's Field Guide (Niehaus & Ripper 1976) and Hardy's (1964) Wildflowers of the Pacific Northwest. Another Peterson Field Guide series was used for the Rocky Mountains and mid-continental region (Craighead et al. 1963). Bill's botanical expertise was of limited use in off-season travel to Europe, and it was hampered by other interests to attend to. Furthermore, there was a lack of field guides at the time of travel. Other than a delightful trip in 1968 to the Abisko region of Sweden in prime flower season, there are only snippets of botanical identifications made elsewhere, unlike what prevailed for his birding observations. From these sources lists of botanical observations were compiled for three North American regions as well as for selected areas within each where the intensity of collection and identification was repeatedly sustained. Another compilation was made for Eurasia, but the data was too scant to warrant a list for other continents. The lists, about 50 pages all told, place the species into their appropriate family which, in turn, are arranged in phylogenetic order.

Finally, in these diaries, the story slowly unfolds as to how Bill did not become a botanist by profession. However, he was a botanist for life and at 6 weeks before final passage he would leave the rest home to identify flora on the nearby beach of the east coast of Vancouver Island, where he resided. In fact, his last 'ID' was a saline-tolerant beach species which he had never seen before. It made his day, and it was the last identification of any life form new to him before passing away 6 or 7 weeks later.

Botanical investigations through the years

Table 1 summarizes the botanical effort during Bill's student academic years. Because summers were directed to integrated study of waterways, it is interesting to note the broad array of biota identified in these field surveys, which reflects the diverse abilities that Bill had. The table shows his taxonomic prowess with all life forms in the riparian zone, although the birds and fish are omitted because they are noted elsewhere. Taxa is used here, because not all organisms were identified to species level; some were keyed to the generic level, and many taxa of algae and arthropods were field identified only to the family level. For the

Table 1. Numbers of taxa identified by W.E. Ricker in his student years (1923–1937).

Year	Location (Ontario, BC)	Numbers of taxa of biota[a]				
		Flora–terrestrial, aquatic, riparian	Riparian insect, arachnid myriapods[b]	Other F/W aquatic invertebrates	Amphibians and reptiles	Riparian mammals
1923	North Bay	10	N.D.	N.D.	N.D.	N.D.
1924	North Bay	19	1	2	5	14
	Plattsville	19				
1925	North Bay	52	12	1	8	24
	Plattsville	56				
1926	North Bay	98[c]	14	0	10	13
	Plattsville	111[c]				
1927	North Bay	42[c]	0	0	0	N.D.
	S. Ontario	390[c]	7	6	9	8
1928	North Bay	13	1	0	0	N.D.
	S. Ontario	113	137	30 + 10 fossils	10	12
1929	Frank's Bay	166[d]	28	8	7	10
	Algonquin Park	37	27	9	5	14
1930	Mad River	312[d]	196[d]	50[d] + 4 fossils	14	7
	Lake of Bays	15	8	1	0	4
	Gr. Toronto	21	21	5	5	4
1931	N. Ontario	61	135	10	14	25
	S. Ontario	82				
1931–1937	Cultus Lake, BC & vicinity	313	84	43	8	14

N.D. = none detected from journal records.
[a]Taxa = species or genera where species not ID, or family where genera not ID, fish and birds not enumerated for this review.
[b]Based on field identifications, mainly: samples collected for lab identification not available.
[c]In these 2 years plants were collected, and 354 species placed in herbarium at University of Toronto.
[d]Counts are higher than stated elsewhere (e.g., Birding Career) because of more intensive scrutiny in Bill Ricker's diaries.

higher orders of flora, however, Bill attempted to identify to species level but in some areas where the field guides are out of range or incomplete, as much as 10% of the total taxa were identified to generic level only. From the aquatic-based summer work one can ask: when did Bill begin his interest in the taxonomy of stoneflies? Surprisingly, it was not as early as one would perceive. In Table 1 there are no stonefly identifications to species level until 1934. When back at University of Toronto for some Ph.D. winter course work in that year, Dr. E.M. Walker and colleague, Fred Ide, suggested he tackle this order of insects. Up to that time he had been identifying Ephemoroptera, Odonata and some Trichoptera to species level, for some of their genera, but stoneflies were recorded in the sampling work as an order, or sometimes to family, but rarely to generic level. Identification of all organisms in the riparian and aquatic habitat is key to understanding the food pyramid relationships of a trout's diet and the incremental stages in their growth cycle. In fact, before he finished his B.A. he was involved in controversy as to where hatchery fish should be released according to their size and hence their food requirements, and their need for camouflage or a barrier to escape predators. A long public debate on the issue was carried through several newspaper editions of the Toronto Star and rival Globe and Mail.

High school years (1923–1926)

High school botanical interest was triggered by his older sister, Evelyn, who had taken a botany course, and hence a start by them jointly to collect specimens, beginning on 11 June 1925. Up to that time all species identifications were by common name only; fortunately, it was the name listed in Gray's Manual of Botany. Table 1 shows that the number of species recognized at Plattsville was about the same number as at North Bay. Comparing the lists for each locality, however, reveals a significant difference in floral assemblage as shown, for example, by conifers up north and hardwoods to the south. The collecting of botanical specimens in 1926 was double that of the previous years, due in part to the botany course he had taken at high school. His favourite specimens appeared to be *Lycopodium*, identifying five species of the club moss in that year.

Over the course of the year he made his own pair of skis from maple, steamed to produce the correct curve at the tip and the correct camber beneath the binding. Those skis were used for his Mad River thesis work and irregularly until the early 1950s; unfortunately they were not archived as a museum piece. Before enrolling at university he read Arthur Conan Doyle's ('The Parasite'); and in later years he became a fan of the Baker Street entourage. Of this author's Sherlockian characters, Bill's favorite was Moriarity with whom he defended in his own pastiche, written some 60 years later. In 1926 he also began the identification of dragonflies to species level, an ability put to use during his following years as a summer field assistant.

University of Toronto (1926–1927) and fisheries work at Lake Simcoe and Port Credit in the summer of 1927

Bill entered the University of Toronto, Victoria College, with no personal aplomb; the dazzling marks from the matriculation exams were already forgotten, though not likely by the professors he was to have in his first year. Was he a marked student? Initially, probably not, though 'Doc' Satterly (physics, calculus) and Prof. Thompson (botany) were Bill's favourite instructors and there was much friendly communication with them. The most boring professor of the lot, however, was Dr. Kendrick (chemistry) who took over 3 weeks 'to define a saturated solution', as bemoaned in his diary. He was not a night and day student, however; evening study was rarely in his program because he had many other interests. He joined the university orchestra as a second violin player; was an active participant in a choral group; was a performer in a major stage play, and ushered at another big production over several showings. On Saturdays he attended varsity football games and went to church with his older sister at some opportune time of the day on Sundays. Living in a dorm on campus there were the usual pranks, and other distractions. His diary for the school year is full of notes on after-school events and it also noted that he skipped many of his Scientific German classes on Friday mornings!

Despite the antics, outside and other on-campus interests, Bill passed all of the term ending exams in mid-January. In fact, Dr. Thompson told him ahead of time that the mark in Botany was 100%. Soon the word went around from professor to professor; there is a new student who hardly studies because of his so many other activities, and yet he 'aces' the exams and completes his laboratory work to a standard above the usual student level. From the writer's experience in his academic years, it goes without saying that this is the type of student that professors yearn to find for postgraduate potential: comprehends easily in the classroom; 'plays' hard outside of the classroom; yet sails through the course with ease.

In the second term the introductory biology course was begun under the tutelage of Professor W.H.K. Harkness. The labs for the course provided the usual glimpse of a few specimens of the animal kingdom, beginning with the amoeba, progressing to the skull of a cat, and finishing with a study of the embryological development of a salamander egg. It was during this latter exercise that Bill and two classmates were called into the professor's office; would they like a summer job working with fish? For Bill it had not been his first attempt at finding a summer job, for in January he had seen an advertisement on the notice board in the botany building, seeking a summer student to assist in botanical collections for the Province of New Brunswick. Enquiring with Dr. R.B. Thompson about the position, he regretfully explained that it had already been taken by an older student, but he promised to try and find something else for Bill. By 21 April, however, there were still no summer employment opportunities in botany, and so he quickly accepted Dr. Harkness' offer on the following day.

Nonetheless, Bill's botanical career had not been stymied at that point. He was the top student in the botany class, obtaining a first class at year-end, while a new friend, Roy Cain, received a second. Dr. Thompson added some prerequisites for the second year course in botany, which was a survey of the spermatophytes. A collection of dried and pressed specimens, identified to species where possible, had to be assembled as part of the course requirement. The student who had the most species collected and correctly identified would receive a $10 prize. Bill already had 97 specimens in his plant press from 1926, that he and sister Evelyn had collected. The field season for the fisheries work began at Lake Simcoe; about 13 species of plants were field identified of those collected there. Moving to Port Credit, Ontario 2 weeks later, much of his spare time was put to collecting and drying out more plants, stored in the press, and 45 of the many species collected were identified in the field. A brief mid-summer holiday to Plattsville added more plants, but at season-end in mid-September he was still not happy with the quantity. At home in North Bay for 3 weeks in September, the pace of collection was re-doubled, netting 42 easy field identifications of the many collected which, for the summer, totalled up to 297 species that had not been collected in 1926. All told by combining the two collections, there were 354 species submitted, which excluded the upper cryptogams because they were not to be part of the submitted flora. Many of the species on the final list (1927) were aquatic, taken during field fisheries work, and the list also had 44 species of grasses, sedges and rushes which are not the easiest flora to identify. Of all specimens collected he out-did it with the pond weed, *Potamogeton*, collecting 10 species and identifying at least two more in the field. On other fronts it was a busy summer; he listed 88 species of birds and 62 fish species at Port Credit. Moreover, with the specimens of cyprinids on hand (53 species locally) he was able to work out a key to their identification. It had been a landmark summer on several fronts and certainly his acquisition of diverse biological abilities were coming to the forefront.

University of Toronto (1927–1928) and summer work on private trout ponds and streams, northwest of Toronto

The second year at university was 'heavy', and the only courses without labs were English and German. Two courses each in chemistry and physics were the time consumers and at that stage in Bill's life he appeared to be focusing on a career in the physical sciences because he had also registered for a second course in geology. The biological courses were not easy either, but provided relief. Dr. R.B. Thompson was the botany professor once again with his course on spermatophytes. For zoology it was invertebrates by Dr. E.M. Walker, an outstanding entomologist. Labs for the latter tended to focus on local fresh water fauna, which greatly assisted recognition of stomach contents in fishes that Bill was to examine the following summer. Bill's diary for the first term was again filled with outside activities which included memberships in the Biology Club and an outside natural history organization, the Brodie Club, jointly organized by staff of the Royal Ontario Museum and the University of Toronto. After a 2 week small pox quarantine, enforced by the University because he had never been vaccinated, and there were several cases in the city, he gave a talk to the Brodie Club on the birds of Port Credit.

In the second term, the reckoning of the 1926–1927 summer plant collections took place. Carefully, Dr. Thompson inspected each collection; Bill's identifications of the *Aster* genus were deemed to be optomistic and all were down-graded to *Aster* spp.; otherwise the collection was fine, but he lost the contest on quantity. Roy Cain had over 600 species in his plant presses! Exams a few days later, however, netted first class marks for all(?) courses and 100% in geology from Dr. Parks!

For the summer, Dr. Harkness directed the Ontario Fisheries Research Lab to a new series of studies that were funded by private trout farm clubs, and other ownerships. Farms at Englewood (Caledon Mountain Trout Club), Shelburne, Hornings Mill, Glen Major, and on the Mad River, all north-west of Toronto, required a complete ecological study, from the physical properties of the water and terrain itself to all multicellular biota in and adjacent to the waterway. When ecology became 'of age', it certainly began on these projects. Diets of the fish, all species, had to be inspected during the surveys and hence the ability to identify all forms of biota ingested and partially macerated was needed. Reports on the facility under study had to be presented to the owners over the summer. The summer's work began with transportation hassles, and Bill promised Dr. Harkness that he would solve it by buying his own car. After a few sessions at used car lots, a month later he did buy one, a Ford Model T; it had four flat tires(!) during the course of the day of driving it to the scene of field work and flat tires among other breakdowns plagued their field movements throughout the summer.

Botanically, the summer produced 126 identifications in his diary and field books. Each investigation of a trout pond featured a simple vegetation map of the pond and its surroundings (Figure 1a–c) sketched out in his field book. Following a short 2 week study on the Mad River, they had a rest at Collingwood. The so-called ice cave in the escarpment above and south of town was visited, netting species of ferns he had not seen before, but the object was to find a *Grylloblatta* at the cave (no luck). However, it was a fine summer season for identifying aquatic fauna. Many dragonfly species were spotted and collected, including a prized *Gomphus scudderii*! A stonefly nymph (family not stated) provided his first recognition of that order of insects on 27 May, and a generic recognition of a *Perla* species came later at Mad River. As in 1927 the field work concluded in early September and Bill went home to North Bay for 2 weeks before returning to the University of Toronto.

University of Toronto (1928–1929) and summer work at Frank's Bay and Algonquin Park

University life for the third year was more demanding than ever; Dr. R.B. Thompson offered him a position as a first-year botany lab demonstrator, but wily Dr. Harkness butted in with a reminder that there were many trout stomachs from last summer's work that remained to be analyzed. He offered Bill $75.00 to complete the work. It was not clear whether this amount was included with the $482 earned during the summer which, along with $62.00 left over from second year, would be his projected $544.00 budget for the entire school year. However, the Angler's Club provided him with a scholarship to ease the financial pangs. So, Dr. Thompson was outmanoeuvred once again.

Regardless, the schedule of extra-curricular events seemed to grow with each year of academic life. His diary relates on Biology and Brodie Club meetings, several nights at operas, training for an autumn track meet, varsity football games, meetings of the Toronto Angler's Club, seminars, two weekend visits to trout ponds that had been surveyed during the summer, a girlfriend who needed scholastic tutoring, movies, house (dorm) parties with their usual stunts, the Songsters (campus choral group), squash and card games. He retired from playing his violin in the campus orchestra. The Brodie Club had him give a talk on the 'Geology of Ontario' in the first week back on campus.

(a)

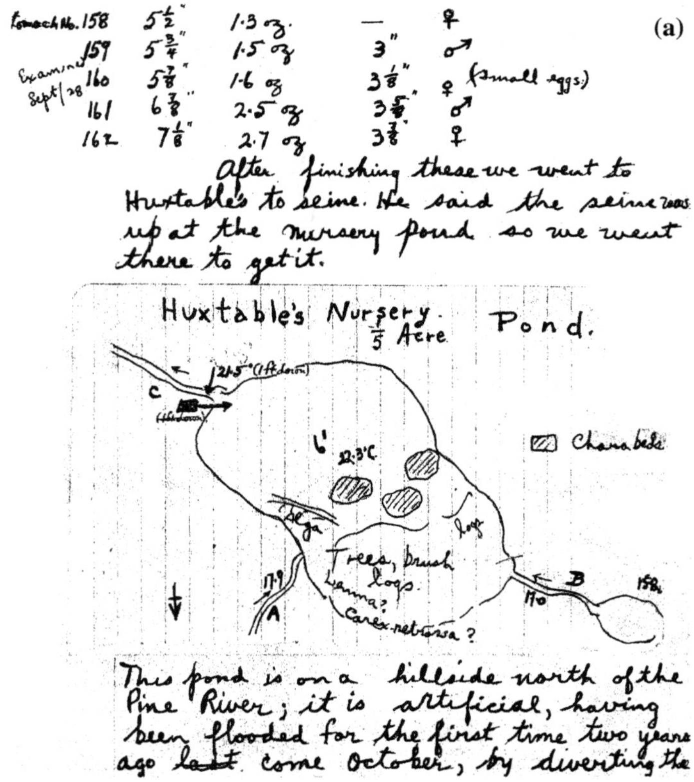

Stomach No.	158	$5\frac{1}{2}$"	1.3 oz.	—	♀
	159	$5\frac{3}{4}$"	1.5 oz	3"	♂
Examined Sept/28	160	$5\frac{7}{8}$"	1.6 oz	$3\frac{1}{8}$"	♀ (small eggs)
	161	$6\frac{7}{8}$"	2.5 oz	$3\frac{5}{8}$"	♂
	162	$7\frac{1}{8}$"	2.7 oz	$3\frac{7}{8}$"	♀

After finishing these we went to Huxtable's to seine. He said the seine was up at the nursery pond so we went there to get it.

This pond is on a hillside north of the Pine River; it is artificial, having been flooded for the first time two years ago ~~last~~ come October, by diverting the

(b)

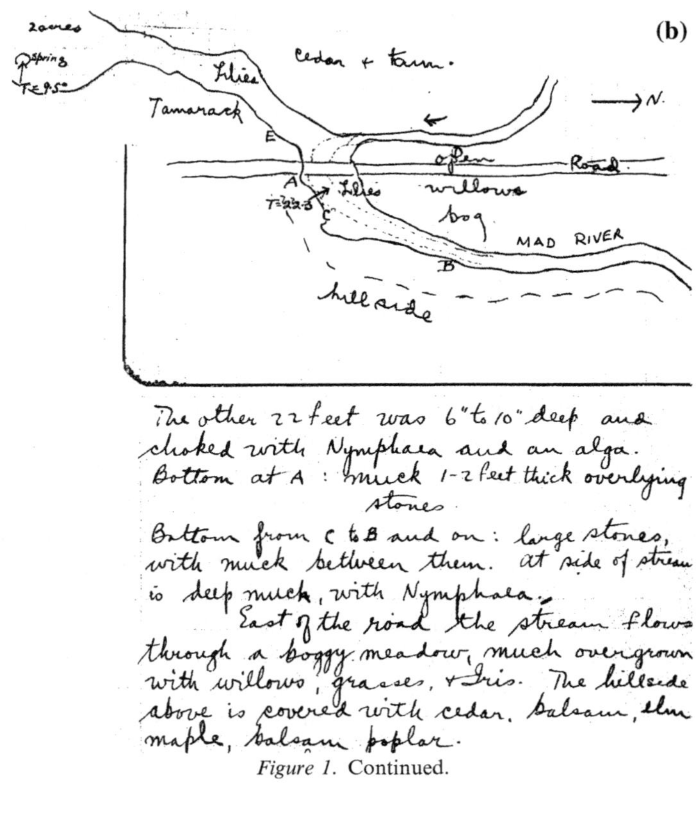

The other 22 feet was 6" to 10" deep and choked with Nymphaea and an alga.
Bottom at A : muck 1-2 feet thick overlying stones.
Bottom from C to B and on: large stones, with muck between them. At side of stream is deep muck, with Nymphaea.
East of the road the stream flows through a boggy meadow, much overgrown with willows, grasses, + Iris. The hillside above is covered with cedar, balsam, elm, maple, balsam poplar.

Figure 1. Continued.

(c)

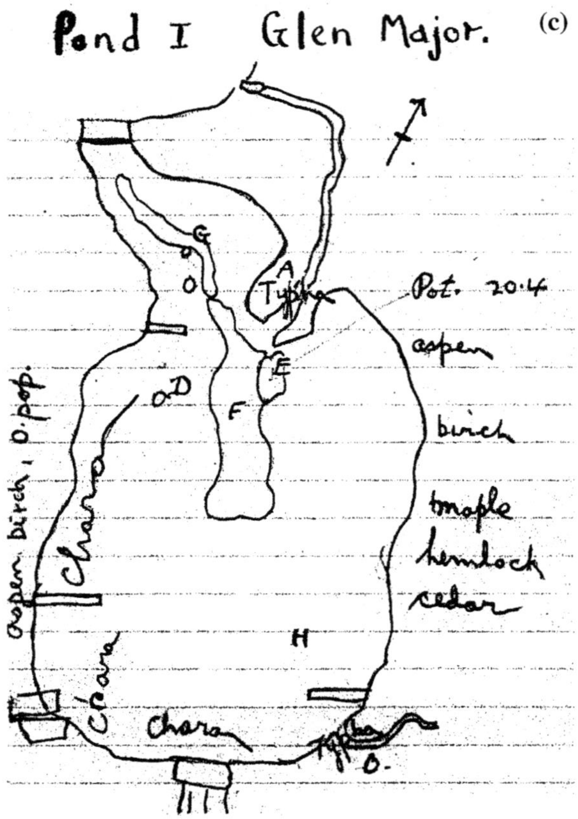

Figure 1. (a) Field book sketch of vegetation distribution at Pond No. 3, Caledon Club, Ontario; from diary/field book of W.E. Ricker for 1928, Vol. 2: p. 16. (b) Field book sketch of sampling stations and vegetation distribution at Glen Major, Ontario; from diary/field book of W.E. Ricker for 1928. (c) Field sketch of vegetation distribution and water quality stations, with notes on fish fauna of Chalk Lake, Ontario; from diary/field book of W.E. Ricker for 1928.

The courses for third year were time-consuming though enlightened by a non-science elective – Psychology. The favourite course for the autumn session was Plant Physiology given by an admired Dr. Duff. In his class there was an assigned essay on 'The Effect of Intensity of Illumination and Concentration of Carbon Dioxide on the Rate of Photosynthesis in Green Plants' which garnered Bill a first class mark. In those days such assignments were hand-written, neatly in fluid ink; Bill used 12 references, four in German, to set up the essay. Within it there is liberal use of differential calculus to explain rates of assimilation. His adroit use of mathematics to explain biological phenomena had appeared. Other courses in the first term were Biochemistry, Embryology, and Comparative Anatomy.

The second term for the third year was postponed for a week by the university; because a flu epidemic was generating chaos in Toronto. The evenings were again booked up, as above, when he did return, but NHL hockey replaced the varsity football. Lower Cryptogams in the Botany Department was a new course which required microscopic lab work and heady book reviews. Flipping through his lab book (Figure 2a, b) it becomes obvious on how Bill managed to identify aquatic algae in the summer surveys that were to come. During the term Bill began to respond to columns in the Toronto Daily Star on the stocking of hatchery trout of varying sizes according to the available food and shelter in the water column. This controversy carried on for several years, using newspapers as the forum.

At the end of the school term Dr. Harkness hired an ever-increasing legion of student assistants for a new project at isolated Frank's Bay on the south-east end of Lake Nipissing. It was to be a full analysis of all components in the ecosystem of the area. Insects were trapped and identified and by now Bill was not only identifying dragon and mayflies with relative ease, both adult and larvae forms, but also was learning the systematics of the caddis flies (Trichoptera). Stoneflies, however, were still being shunned. With the fishes, captured in a variety of gill nets, seines and by hooks (about 35 species) the stomachs of all were analyzed, and a few water birds that had been collected were also subject to gut analyses.

While several students worked with Bill on the above tasks, he and another also ran a small mammal trap line and kept records on the sightings of bigger species which were few. Any animal of muskrat size or smaller was skinned, and then stuffed for submission to the Royal Ontario Museum. Over and above these duties, the vegetation was identified, and collected where significant, and the avian fauna was recorded. His list of flora for the Frank's Bay region (1929) registered 158 species, although his diary indicates a few more (Table 1). The prized specimen for the area was a black ash, *Fraxinus nigra*, although Morton & Lewis (1921) show a broad distribution of the species throughout eastern Canada. By the end of July the Frank's Bay project was going smoothly. Drs. W.H.K. Harkness and J.R. Dymond had another project in mind, also on the Canadian Shield, and sent their most experienced students to work on it alone.

A canoe trip through the Algonquin Park region is what every young man dreams of, and now Bill with Fred Ide were being paid to do it! Because of capacity limitations in a too short 4-m long birch bark canoe, there was no possibility of collecting botanical specimens on this journey. Fish stomachs were the priority cargo, over and above what they needed for living, but some insects were also collected and stored in small vials. His hard-covered diary was also in the packsack, along with a bottle of ink to record the daily observations. But alas, the wooden stick pen fell into their camp fire by accident and the remainder of the journey was recorded in pencil. The diary notes 37 species of plants on their canoe trip, but it also has thoughtful notes on forest succession in the northern part of the park where fires and logging at various times have created a 'brule' now in various stages of re-colonization. Untouched *Pinus strobus* along parts of the Nipissing River were a welcomed sight in the final days of their journey. Notable identifications while working and portaging in the riparian zones were five species of *Lycopodium* (his boyhood favourite) and six species in the family Orchidaceae. The canoe trip was completed in late August, and after attending a conference in Toronto Bill returned to Algonquin for another week alone in early September. He told me (as a teenager) that the solo trip was not a smart decision, but apparently he and Fred had some disagreements on sampling areas in their August journey and Bill was determined to fill in the gaps. Bill was trying to pass on a lesson to a solo wandering son(!) and although the advice was not always heeded it was always in my mind whenever I went into the wilderness alone.

Fourth and graduation year at the University of Toronto (1929–1930)

The fourth academic year, first term, was 100% biological, no courses in other departments, although the Botany and Biology Departments were separate entities. He took three courses from the former and four from the latter, plus the compulsory 'History of Biology' for all honours students in both departments. From the Biology Department the heavy course was Systematic Zoology with two labs per week. Unfortunately, neither course notes nor laboratory drawings could be found to reveal its content but it was probably taxonomic as well as evolutionary. Equally unknown is the content of a course in Hydrobiology, given at 17:00–18:00 hrs on 2 days per week, and it was without a laboratory component. The Histology and Genetics courses rounded out the Biological Department's offerings, but the latter course obviously had a direct relationship to the botanical field.

From the Botany Department, Upper Cryptogams from Professor Jackson was the time-consuming course; Botany Seminar at 1 hour per week was the light one, and Plant Ecology (by Dr. Sifton) was the course with field laboratory on some Saturdays. Unfortunately, his notes and

(a)

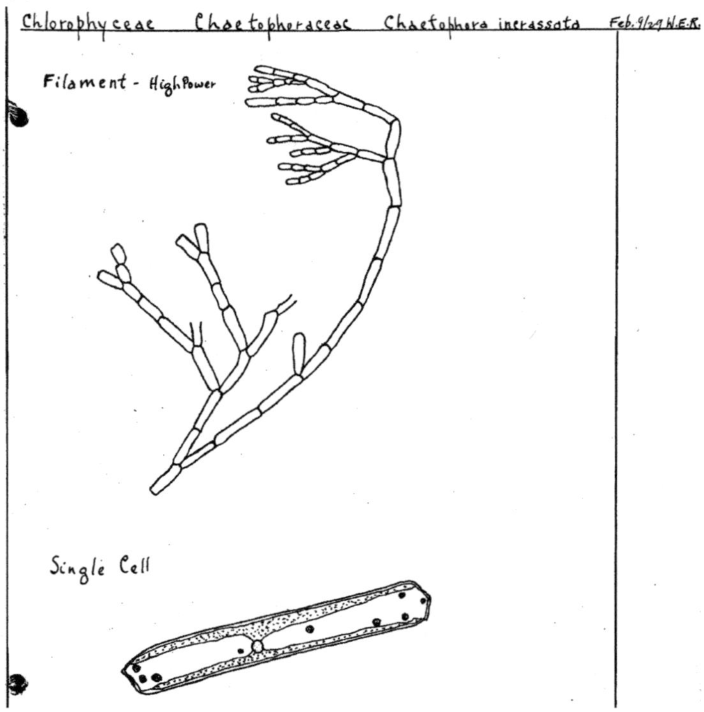

Figure 2. (a) Drawing from Lower Cryptogam Course (1929b) – Chlorophyceae, Chaetophoraceae, *Chaetophora incrassota* (b) Drawing from Lower Cryptogam Course (1929b) – Charales, Characeae, *Chara* sp.

laboratory exercises for this course have also been misplaced, although not his essay which had a direct bearing on his summer aquatic studies of fish habitat (1930). The cryptogam course began with Hepaticae (liverworts), progressed to Bryophyta (mosses) and finished with the Pteridophyta. Some of the drawings produced in the laboratory are a work of art (Figure 3a, b) and obviously prepared him to easily identify the Pteridophytes in his field fishery projects. Mosses, however, like fungi and moulds, were not his forté as indicated by a lack of recognition of many on his field lists.

Finally, the staff made sure that there was little spare time for Bill during the course of a university class day. Not only was he a lab demonstrator for comparative anatomy on 3 days per week, but also they placed a mandatory attendance and participation at their 'Biological Journal Club'. It is not clear whether there was course credit for the latter, but during the autumn session Bill was asked to prepare a talk on the following topic: 'Whereas generic differences are usually of adaptive significance, those which distinguish species are not so'. 'Oh horrors', he wrote in his diary and the talk had to be delivered early in the following winter term.

Well, the onslaught of academia was not about to slow down Bill's other pursuits in the evenings or on the weekends. Nearly all out of class hours time in the first term was taken with squash and bridge games (with his professors yet!), class parties, theatre, show, musicals, a trip to Niagara Falls, inspection of trout ponds at nearby Acton, more fish stomach analyses for Dr. Harkness, the Brodie Club, and a new twist – he was the elected president of the student Biology Club. And then he did some relaxing reading, if 'Tarzan of the Apes' does reduce the stress of a fully-committed student. On one hectic weekend he went so far as to skip an ecology field trip which was part of a laboratory assignment.

In the second or winter–spring term the pace was not reduced. The cryptogam course had all but wound up in January, but Professor R.B.

(b)

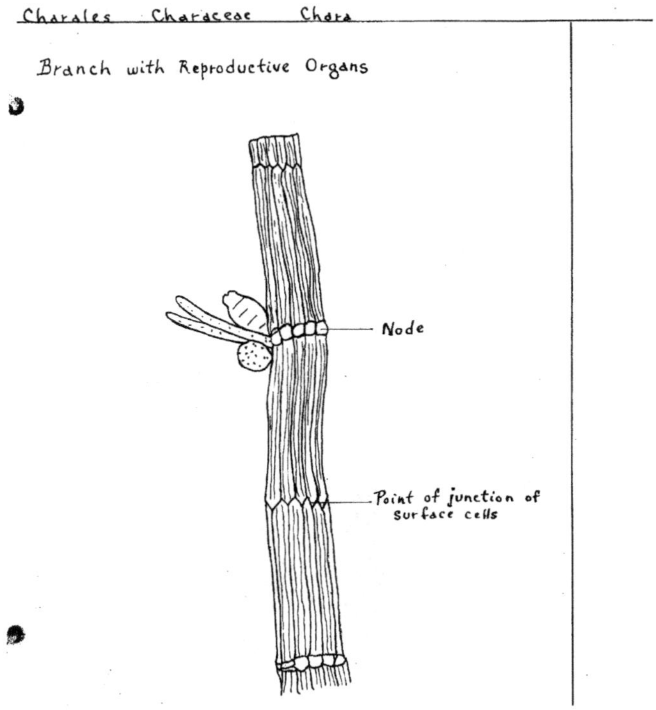

Figure 2. Continued.

Thompson, now the Department Head, assigned a lab demonstrator's job to him on Saturday mornings – introductory botany for engineers. Bill was not impressed with their work ethic, and marking their drawings each week provided a class average that was a consistent C minus! There were no other course changes from the first term, but the cryptogamic botany course had two outstanding items – the submission of his collection of cryptogams (no records located) to Dr. Jackson and an essay which was finally finished on 31 March 1930: 'The reduction of the gametophyte and amplification and sterilization of the sporophyte in the Archegoniatae'. As before, it was written out in ink format which contained text surrounded by neat pencil drawings (18 sets altogether), as shown in Figure 4. The talk to the Biological Journal Club on 'adaptive generic differences' was well-received and Professor J.R. Dymond urged that he publish it, which he did in the Canadian Field Naturalist (Ricker 1932), but with an altered title.

In the after hours the scene became very hectic because the Biology Club chose to have a campus open house on the topic of demonstrating various aspects of laboratory and field work in biology, entitled with posters plastered all over the campus 'Conversazione'. Bill was the coordinator of the event which took place on 19 February 1930. And on Easter weekend (April 17–20) the club had a field camp at Kelly Lake, north of the city. Fitted around these two commitments were more of the above evening events, added to with graduation exercises, seminars by specialists visiting the university and his swimmer's Life Saving exam. The 7 March 1930 edition of the local varsity newspaper ran a short column spoof on the demise of Bill Ricker, announcing that he had just finished describing his 56th new species of water cress and his partner in botany, Roy Cain, was suffering from suicidal tendencies. Well, Roy stuck with botany to eventually become a staff member of the Department at Toronto, and Bill was actually writing an essay for the ecology course at the time entitled: 'A Comparison of the *Potamogetons* Living in Swift Waters with Those Growing in Ponds'. At that point in time he had identified 15

(a)

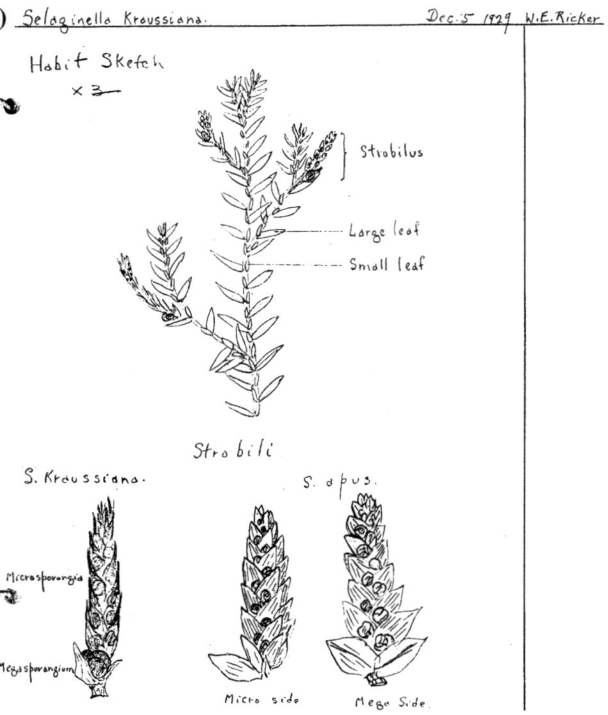

Figure 3. (a) Drawing from Upper Cryptogam Course (1929c) – *Selaginella Kraussiana* and *S. apus.* (b) Drawing from Upper Cryptogam Course (1929c) – *Lycopodium clavatum and L. lucidulum.*

species in Ontario, which is impressive, for most taxonomists stop at the generic level for the identification of many aquatic monocots. Certainly at this stage in his career the recognition of the species of aquatic flora was his forté. The essay, another first class effort, is again hand written ink script on unlined paper; there are 41 carefully prepared sets of diagrams, in pencil (Figure 5, for an example), and two of the key references used are in German.

Despite the continual interruptions in the academic course work Bill finished his graduation year in fine style. The university transcript for student grades was published showing course grades for all graduates; Bill received first class marks in biology and botany, ranking number one among the students graduating from the two departments. The only 'hitch' was that he received a 'B' in the History of Biology course although all other students were given a 'C'. Among all science students his rival was the later famous J. Tuzo Wilson who was number one and first class in geology and physics, but he was given a 'C' in Religious Knowledge! No wonder, since plate tectonics and religion can hardly be a compatible train of intellectual thought. So when it came to parcelling out the awards the senate decided that Bill would receive the G.A. Cox Gold Medal in Science and J. Tuzo would receive the A.P. Coleman Gold Medal in Geology. While the latter also nabbed a Governor General's Silver Medal, Bill was awarded the Bronze from the British Association for the Advancement of Science. So the net result between the two was a standoff, and as for Bill's rival companion in Botany (Roy Cain) he wound up first class as well, but number two in the ranking.

The M.A. thesis project at Mad River (1930–1931)

All of the star students were continuing to graduate studies. For Bill's master's degree the Ontario Fisheries Research Laboratory had financially backed his proposed project for the study of the Mad River. In his proposal he set out the following goals:

(b)

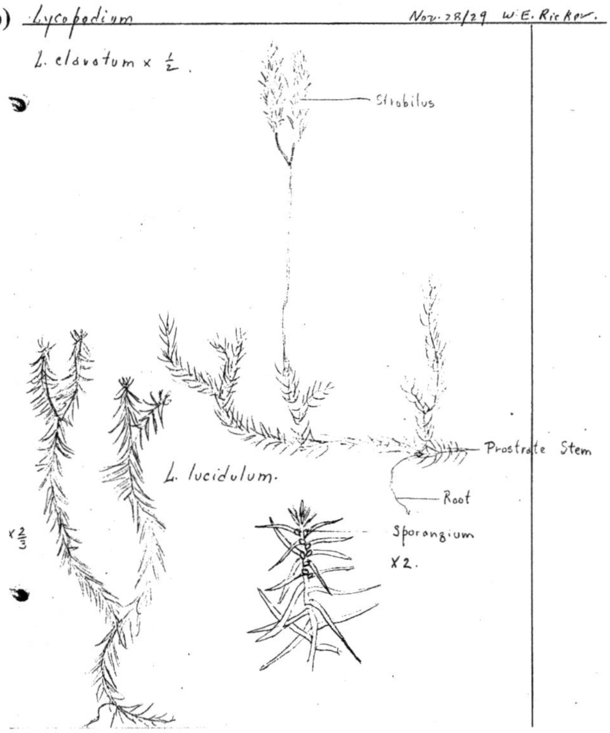

Figure 3. Continued.

(1) at four or five stations to be set up on slower moving reaches of the river, located up-stream of the Niagara Escarpment, measure physical properties of the water, including discharge rates;
(2) collection of aquatic biota, noting times of their emergences and/or spawnings (the terrestrial portion of the riparian zone was part of this exercise);
(3) quantitative studies of food supplies for trout and stomach analyses to verify their utilization; and
(4) tagging of trout to ascertain rate of growth in wild conditions.

The stations were to be visited on weekly to bi-weekly intervals throughout the study period which was proposed to conclude in the spring of 1931, that is, a full year of survey. From the Mad River survey he proposed to expand the analysis by comparison to non-trout producing waters elsewhere in the province and to come up with a classification of trout-bearing waters, noting the distribution of each within the province (Ricker 1931, 1932a, 1934).

There were also other plans for the summer, including a return to Frank's Bay to help out Professor Dymond. However, Dr. W.H.K. Harkness added a few other projects. As usual, the summer program required a car. A trip to Plattsville to retrieve the Model 'T' stored at his grandparents' farm came to the conclusion, after brushing off the dust, that one in better condition was needed. So he decided to trade up for one of the same model but in better shape. The diary lists the various components of each car, his and the one available, rating the condition of each. Frankly there was little difference except the replacement vehicle had a battery assisted starter as opposed to the manual crank, and the emergency brake actually worked! Driving the vehicle to Toronto he had a collision with another vehicle, breaking a headlight and puncturing the radiator, and the engine misfired all the way, despite several stops to adjust the timing. On 14 May he was on his way to Singhampton (the base of operation on

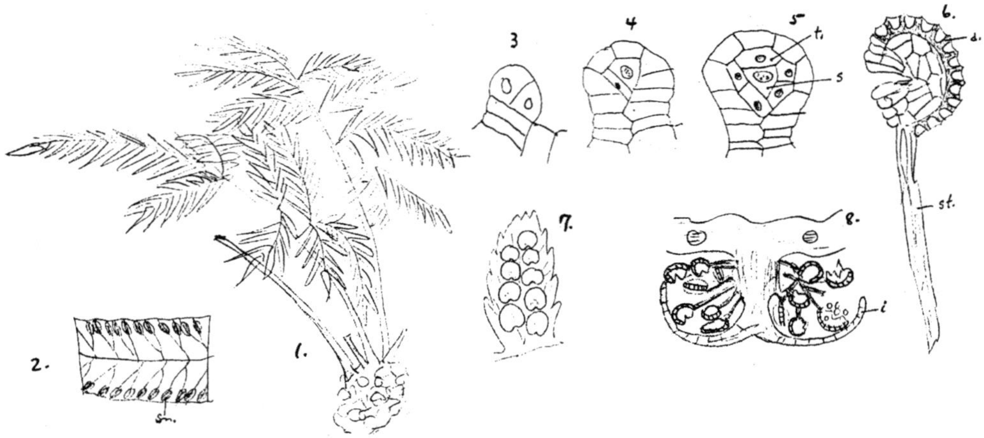

Figure 4. Sporophyle structure of Filicineae, from essay by W.E. Ricker, Upper Cryptogam botany course, 1930e, Univ. Toronto. (1) *Morattia*, (2) lower surface of leaf of *Morattia*, showing synangia, (3 – 6) development of sporangium of *Aspidum*, (7) Sori of *Aspidium*, (8) cross-section of sorus, a – annulus, st – stalk, sn – synangium, i – indusium, t – tapetum, s – spore mothercell initial.

the Mad River), 83 miles (134 km) north-west of Toronto at the average speed of 39 km h^{-1}, noting that the oil consumption was high.

The Mad River project had much biota to identify and measure, which kept him busy throughout the summer. A list of botanical specimens identified, with some collected, came to 293 species (Ricker 1930) but upon reading his diary several other species could be added to the list (Table 1) based on excursions away from the river. The invertebrate collections were also significantly diverse in species or taxa, but by this time Bill had the skills to identify almost any form of biota. Once the initial field stations were established and the various parameters of the project had been measured or collected, Bill spent a week with Roy Cain in early July on streams to the west of Collingwood near the Bruce Peninsula of Lake Huron, also noting unusual beach flora at lake edge – definitely a different floral assemblage in that part of the province.

A new project appeared after the visit to the Bruce. Dr. Harkness had asked him to investigate potential trout rearing sites in a region known as the 'Lake of Bays', centred on Tally Ho Creek and the Oxtongue and East rivers, near Huntsville. The survey did not provide a suitable rearing site much to the dismay of the proponents, as shown by a very short list of aquatic plants and invertebrates (Table 1). For the last 10 days of July Bill was back to Mad River to survey his stations again, and again in mid-August after a trip to Plattsville. At month-end he attended the American Fishery Society (AFS) meetings at Toronto where he read his first paper on trout feeding habits. The proposed Frank's Bay visit had been eliminated by the other assigned activity and meetings. After the Labour Day weekend the Mad River stations were re-surveyed, after which it was decided to take a holiday. With a friend, a 3200 km long car trip led through northeastern United States and back to Toronto by way of the St. Lawrence valley in Quebec. The fuel for the trip cost $26.88, including the 27 pints of oil poured into the crankcase!

As far as can be surmised Bill did not take any biological courses at the University in his post-graduate year for an M.A. degree. Vertebrate Paleontology from Dr. Park in the Geology Department is the only course noted in his diaries. However, during the academic year his time was occupied with lab analysis of his Mad River collections and attendance at seminars. The Biology Journal Club had him reiterate his AFS paper on trout feeding habits, later published (Ricker 1930). Curiously, the after-hours activity was not so hectic as in past years, primarily because he was on the road almost every week. There were trips to Mad River, a few days at the Caledon Club ponds, re-survey of Strother's hatchery creek near Orangeville, the Biology Club field retreat to Gull Lake on the holiday weekend in October, and a trip to North Bay in November, and again home for an extended Christmas holiday. At Gull Lake the prized botanical dis-

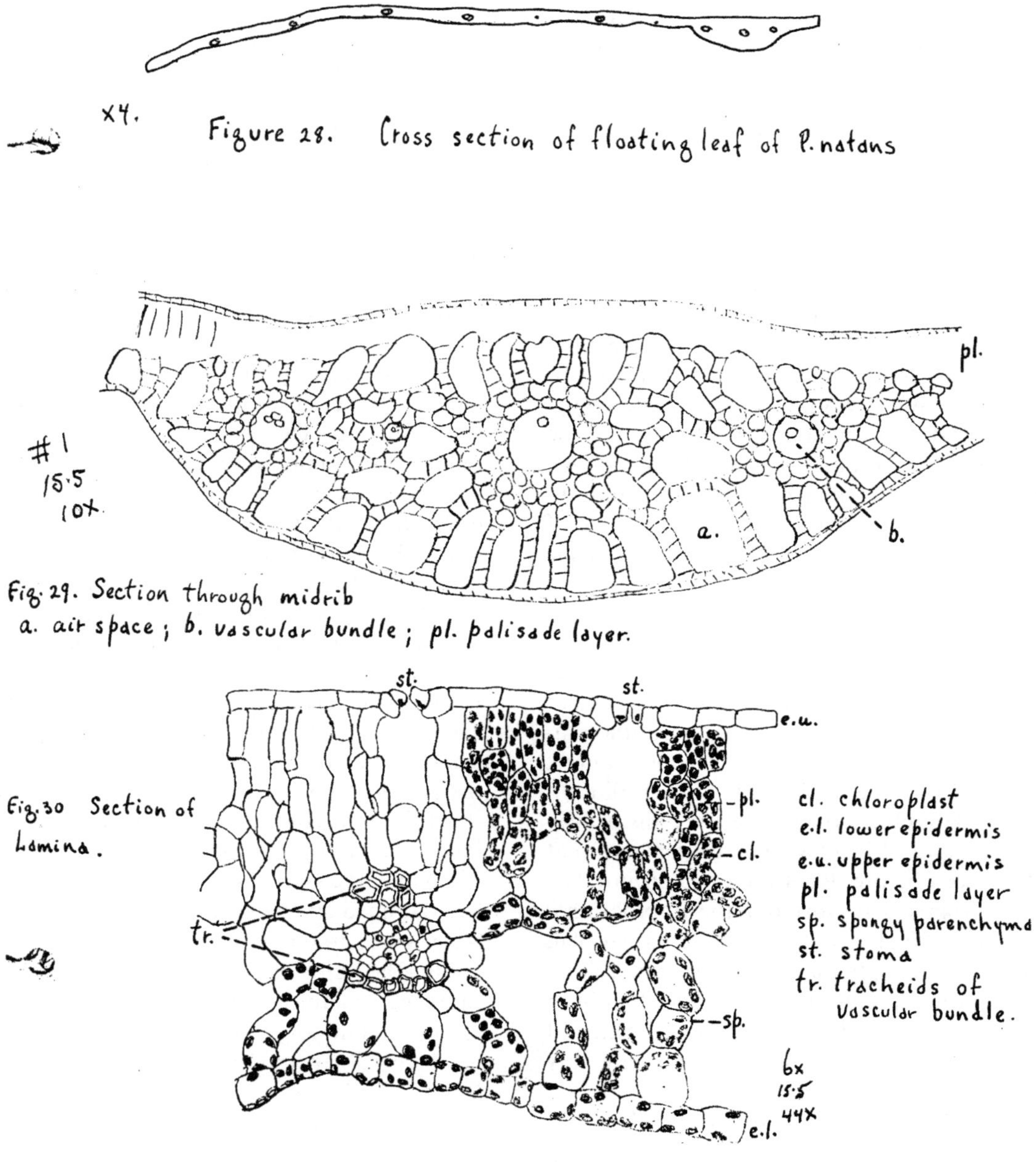

Figure 5. Drawing in essay for Plant Ecology course (1930d) – sections of floating leaf of *Potamogetons natans*.

covery was *Gentiana crinata*, and it was a flower he never saw again throughout his wanderings in eastern Canada.

After the Christmas break there were three more trips to the Mad River that had to be fitted around his commitment to being a lab demonstrator, once again, in comparative anatomy. Surprisingly however, the thesis was wrapped up on 16 April, very quick considering that the last visit to the study site stations was on the third week of March, after spring break up. In late April and early May the Biology Club held two field camps, the first at nearby Kelly Lake and the other at Willow Creek on the Nottawasaga River near Collingwood. Not only did the camps provide some breathing space to think about his future, it also gave him a chance to work with younger students on field identifications. A rare hackberry, *Celtis occidentalis*, an Urticaceae, was the highlight find at Willow Creek.

A summer of wanderings and unknown future (1931)

Returning from the Biology Club camps, the question was what Bill was going to do in the short or long term? In early February Dr. Walker, the Department's entomologist, offered him a staff position at the University conducting experimental projects. He mulled it over but decided to wait to see if any other offers would turn up. Surprisingly, the botanists did not make an offer although he was an obvious candidate for a Ph.D. in ecology. Dr. Harkness on the other hand was scheming to set him up for a Ph.D., working with the State of New York's Conservation Department. While negotiations were in progress he was sent to Port Credit to sample the fishery and at its conclusion he loaded the requisite sampling gear and supplies into his car to begin a project approved by the New York authorities. The mission was stopped at the border, no work visa; the letter of authorization by the State authorities wasn't good enough; yet the U.S. Consulate in Toronto said he did not need a visa. He retreated to Fenellon Falls nearby to study its natural history, including plants that he had not seen before. Returning to Toronto, Dr. Harkness promised to sort out the international immigration difficulty, but as an alternative plan he advised Bill to apply for a studentship at Trinity College, bearing in mind that an earlier attempt to secure a Rhodes Scholarship to Oxford had not worked out. The diaries do not elaborate any further on the matter.

In June while awaiting answers and decisions from several authorities on his future, Bill drew up alternate plans for the summer, and he worked on three papers to be published (Ricker 1932a, 1932b, 1934). He also investigated the trout rearing capability at Ethelmere Lake for a private owner. He made two more trips to Fenellon Falls to study its natural history attributes, and another excursion to Niagara Falls to collect insects and identify unique gorge flora. Not least, there was also the graduation ceremony to receive his Master of Arts – the Master of Science was not yet in vogue at Canadian universities in the early 20th century.

On 1 July Bill was off to Frank's Bay along with several staff members to continue the projects begun in 1929. More flora were identified while sampling the waters for fish and other biota. He watched sawflies strip leaves from a conifer over a period of time, and in his diary he calculated that 200 of these larvae could strip a 2-m high tree of all its needles in only 2 hours! Dragonflies and mayflies, however, continued to dominate his identification of riparian insects. Not one stonefly was mentioned in his diary for the Frank's Bay area. On 23 July, after working 3 weeks at the Bay, Dr. Harkness had a quiet 2-hour discussion with Bill about the possible options for the forthcoming academic year. The project in New York State was nullified by immigration difficulties, but there were several, including Dr. Walker's offer which was still on the table. Bill spent the following 3 days with his old high school pal, Charlie Wager, on a fishing trip along the Ottawa River to think it over.

Returning to Toronto, there was a new horizon. At long last the Research Lab had decided to launch field investigations again in northern Ontario at Lake Temagami, about the mining district of Timmins, and farther northwest to Kapuskasing. That is, the area of investigation would extend to the northern side of the Canadian Shield and onto the edge of the Hudson Bay Lowlands, underlain by Paleozoic carbonates and overlain by vast expanses of silt and fine sand of Glacial Lake Ojibway, which had dammed up against a retreating Laurentide Ice Sheet some 10,000 years ago. He was enthusiastic about the trip, because it also allowed much canoeing on fabled Lake Temagami, and a chance to look at a mining district which was once the wild heyday of northern Ontario's economic enterprise. Bill took Gray's botany manual with him and directly noted within its pages the many new species that he had not seen before. Study at Lake Temagami was of all biota, which included the trapping of small mammals as part of the day's work. After a couple of weeks at the lake they moved on to Haileybury, Cochrane, Timmins and Kapuskasing and other areas to sample the aquatic biota and have a look at the surrounding flora. Near Latchford he spent part of the day analyzing forest succession in an old burn area. On 25 August they returned to Lake Temagami to wind down the program but fortuitously, if not ironically, they met a collecting expedition from the Department of

Botany, University of Toronto, which included Professor Jackson, the cryptogam specialist, and Bill's fellow student, Roy Cain.

After a prolonged weekend at North Bay, Bill was back to Toronto filing away the data and identifying the contents of their collections up north. There was still no word on the several alternative plans that Dr. Harkness had been casting about to find a spot for his future. But Bill had now made up his mind; he was weary of Toronto, after 5 years there, and he wanted a new horizon. So, Dr. Walker's entomology offer was rejected. He left Toronto mid-month and hitchhiked for 2 weeks in southern Ontario to see relatives and to visit the farthest south geographic feature in the province, Pt. Pelee. This time botanical observations were the priority, but he bemoans the fact that Gray's Manual of Botany was not with him. Some plants could not be identified from his repertoire and many potential new species to be added to his lists had to be dismissed. It was the first occasion, however, that he had seen a prickly pear cactus, *Opuntia* sp. The trip, nonetheless, had been a rewarding soul-searching journey and he was ready to return to Toronto to see what options were now available for his future. Arriving late on 29 September the critical message was there in a letter which had been posted by Dr. R.E. Foerster on 21 September 1931 from Chilliwack, British Columbia. The offer was to work as a research assistant on the sockeye salmon program at Cultus Lake with him, employed by the Biological Board of Canada at a salary of $1800 per annum.

September 30 was a day of indecisiveness. Various factions in the Departments of Biology and Botany gave conflicting advice on the offer. Some wanted him to stay at the university or go to another institution nearby, while others promoted the offer. Near the end of the day Bill decided to see the Biology Department head, Dr. Bensley, who was above and beyond those with vested opinions. After listening to the various pros and cons that Bill presented, he said, 'I would advise you to accept the offer'. The following succinct message was to be the turning point in Bill's career, and for the rest of his life:

Key message (telegram) 1 October 1937 to Dr. R.E. Foerster

> Accept position at Cultus Lake on basis outlined in your letter of 21st (Sept) at initial salary of $150/monthly. Stop. Since my stay is to be long should like to remain here if possible until October fifteenth in order to clear up past work. (William E. Ricker)

Bill left Toronto on 9 October, after an exchange of letters with Dr. Foerster, heading toward the Bruce Peninsula area to do some more sampling on the Saugeen River system, a continuation of work of his visit there in July 1930, and from there to North Bay on 13 October. The Bruce area was to be the last area of botanical observations in Ontario for a few years, and the only eastern locality where he had seen shrubby cinquefoil, *Potentilla fructicosa* and the savind, *Juniperus horizontalis*. At North Bay he completed preparations for the move to the west and on 15 October, as promised, he was on the Canadian National Railway train heading west with a stop at Saskatoon to receive some advice on limnological methods from Dr. H. Rawson. This was his first trans-continental train ride of many to come in the decades to follow. As far as he was concerned the rail was the king of travel and he rode them throughout his life on several continents.

Investigations at Cultus Lake, British Columbia (1931–1937)

The west coast was a new environment for Bill, arriving in late October at the start of the rainy season when the Aleutian Low Pressure Cell moves southeast along the coast of British Columbia to spin its counter-clockwise moving-winds from off a warm Pacific Ocean to the northwest. The mountains about Cultus rise to 1500 m, and higher nearby, to trap the moisture-laden clouds. Living alone at a small cabin, provided by the Biological Board, was to be a potentially dreary exercise, but Dr. Foerster was not taking any chances. He whisked Bill to the Pacific Biological Station at Nanaimo before mid-November to do some chemical analyses of water samples, meet new personnel who had an interest in the Cultus Lake work, and carry out library research. Back at Cultus Lake on 18 November the work on its waters and at the counting fences at its outlet proceeded.

In December a surprise arrived in the mail: Volume I of the University of Toronto's Biology Club Journal, which was instigated by Bill in his last 2 years on campus. To this day we are not sure whether Bill made a written contribution to the journal or not; his copy has disappeared, likely during the move to Indiana in 1939 or in the move of his books and papers in 1969 at the Pacific Biological Station at Nanaimo. Suffice to say, botanically it was not the time of year to look around but he quickly identified the new conifers, saw the exotic broom, *Cyticus scoparius*, for the first time in his life, and pondered the identification of a club moss species, *Lycopodium* sp., not seen before.

In mid-December the Aleutian Low backed off in favour of Continental High which pushes polar outflow winds outward through valleys to the coast. The lake froze over and with this event was heralded Bill's investigation of the seasonal circulation in an oligotrophic lake (Ricker 1935). Life was quiet regardless of the noise of the cold winds outside. He splurged and spent over one half of his 10 December pay cheque (his first) on an eight tube radio costing $89.50 (with no sales tax). It could pick up radio broadcasts from far away and listening to symphony orchestras while carrying out laboratory analyses became the modus operandi while at Cultus.

In early January 1932 Bill formulated a work plan for the year. On 6 January he received some Toronto Globe newspaper clippings (25 December 1931 edition) about his trout diet studies in Ontario, written by some unhappy angler. Bill fired off a response to the newspaper forthwith, which triggered more letters to the Globe, all of which were duly pasted into his diary whenever they arrived. Had he chosen botany as a final profession, perhaps he could have avoided public and scientific conflict which was to bother him throughout his life's work. He did not really enjoy the combatants' forum regardless of how polite or hostile it may be, and whenever there was a chance to do so, he would dodge taking part in controversy. However, the move to the west coast would soon exacerbate over and above whatever problems the Ontario fisherman posed.

Besides the noting of copious biological observations in the diaries, the back of each book was also used to record expenses to the last nickel. An amusing entry accompanied by copious calculations in January 1932 was about a monthly electric bill for $1.90, much higher than the previous. Why? Noting the wattage of all the electric devices in the cabin, he discovered that his new radio was the culprit! But by month-end the west coast cloud banks were again bothering him, and in the diary he duly noted only one cloudless day in each of October, November, December and January! In February it wasn't much better, and snowfalls during the month often changed to rain later in the same day. At month-end the local river, the Chilliwack (then called the 'Vedder'), was at flood, and the following month of March was also noted to be a very wet month! This type of winter usually sends easterners back from where they came!

However, the weather improved in the spring and Bill's biological exploration of the Cultus Lake basin was now underway. His diary notes stonefly collections at the lower elevations by early May and by month-end he was hiking up through the forests to ridge tops surrounding the lake basin. On 24 July he had reached peaks on Liumchen Ridge, well above timberline, which was his first encounter with the rich alpine floral zone. The diary notes many new forbs and shrubs not seen before, as shown by compilation of about 85 species found in his notes. Interestingly though, in the list of insects dominated by dragon and mayflies, there is mention of only one stonefly identification, *Alloperla* sp.

What were the other activities at Cultus besides natural history and work on the lake and its outlet stream? Local softball, a first visit to Hell's Gate, his first golf match with his new mentor (Dr. Foerster), and a visit by his parents were some. He also joined a gang of 12 others to help begin construction of a log cabin on Liumchen Ridge near timberline over 3 days in mid-September. In October the valley was besieged with forest fire smoke emanating from a large conflagration. By early November, however, the Aleutian Low had arrived to provide the damper which then produced local flooding, when 91 mm of rain fell on 12 November. The moist cycle ended after a few very wet days in early December with the arrival of polar outflow winds which saw the temperature plunge to minus 24°C but ending a week later with the return to an annoying daily rainfall, lasting to 21 December. Christmas and New Year's was again celebrated at Nanaimo with staff at its Biological Station.

The year of 1933 was much the same as the previous although he was now attending bi-weekly meetings of the local Kinsmen Club which featured lively intellectual debates as well as hosting community volunteer projects. At the conclusion of winter, mountain rambling was back in full swing until Dr. Foerster sent him to canneries at Bellingham repeatedly over the mid-July to early September period to search for Cultus Lake tags on salmon caught by fisherman on the high seas. Botanical observations and other activities were curtailed although he did have an afternoon to identify a few new plants at Anacortes. By now Bill was the owner of a new treatise on west coast flora (Henry 1915) which served him well over the years to come, updated by Eastham (1947) some 14 years later. Species seen on any trip about Cultus Lake were often noted in fine pencil on the appropriate page of the book, rather than in the diary, which on inspection did not have many botanical entries for the year. This philosophy also prevailed for the insects, although his diary notes that a collection of stoneflies taken over the spring and summer was sent to Dr. Claasen at the University of Illinois for identification. In early December the work at Cultus Lake was shut down. Using Cultus Lake studies for the thesis topic the Ph.D. candidacy required he take some courses at the University of Toronto for the winter term of 1934.

On 11 December he departed in his vehicle driving back to Toronto through the United States in tough winter conditions; obviously it was not an opportune time for biological observations. However, it did give him his first glimpse of 'basin and range' topography and the appreciation of the variability of plant communities to be found when traversing from desert valley floors to forested mountain slopes.

At the University of Toronto three mathematics courses were taken, of which two were in statistics because these type of problems had appeared when analyzing his data at Cultus Lake, not only with fish population determinations, but also with the plankton counts taken in the extensive sampling of the lower life forms that make up the food chain for the salmonids, and their predators. It was at Cultus Lake where he had begun the task of identifying microscopic algae in the water column. In Ontario only the macro-sized algae such as *Chara*, *Cladophora*, *Batrachospermum*, and *Nitella* were noted. These of course are chains of cells, linked together to provide ease of visibility in the shallow water regions. In the pelagic regions where fine meshed nets are used to collect the phytoplankters the quantitative aspects needed a strong statistical methodology to evaluate the densities of these life forms. Other conundrums encountered at Cultus were the origin of the kokanee salmon, a fresh water race of the sockeye, *Oncorhynchus nerka*, and another, the distinction of the various runs of the sockeye entering the Fraser River from the ocean at variable intervals throughout the summer and autumn seasons to disperse as distinctive races into their hereditary headwater streams to spawn. Thus, he also signed up for a graduate course in genetics. Again, fishery biology or courses directly related to the science were not taken if they were available. And as a lark he apparently audited a course in glacial geology. There were also 'directed studies' but one was easy; he chose to write up his boyhood bird studies at North Bay, while the other directed study was a step into the unknown. Dr. Walker and his long-term student associate, Fred Ide, suggested he begin a study on the taxonomy of stoneflies. The specimens were housed next door in the Royal Ontario Museum, where he spent much of his spare time. It was a busy session. In fact, for the first time in 12 years he did not make daily diary entries, skipping the time span of 6 January to 29 April 1934. He did write a summary for the period, however, later in the year.

As usual, the after work hours at university were again active in the evenings and on weekends; he was a participant in the on-campus Glee Club, the German Club, the testy Biology Journal Club and the Brodie Club. For the Journal Club he gave an address on Marxism, based on reading some high-powered writings of others, and for the Brodie he had an easy address: 'The Natural History of Cultus Lake'. Outside the campus he often went out with younger sister, Isabel, who was by then a university student; there was Federation of Ontario Naturalists field outing to Guelph and some of his classmates were getting married.

With the course work out of the way in early May 1934 Bill motored back to Cultus Lake with sister Isabel in the navigator's seat, on a route similar to the previous, through the United States. It was an 11 day trip and despite the long distances travelled each day there were observations on the flora, the most notable and significant to Bill was the *Yucca glauca*, Spanish bayonets, and a curious member of the Cactaceae family, *Coryphanta viviparia*. The trip route, however, was north of the really interesting cactus-bearing country, but it whetted his appetite to traverse the 'basin and range' country wherever possible in the future, so done many times up to the final years of his life.

The summer of 1934 and following autumn at Cultus Lake featured much hiking on the ridges about Cultus Lake, adding new sightings of flora and fauna to his ongoing inventory. The highlight appears to have been the discovery of a new species of dragonfly, *Macromia* sp. nov., confirmed by Dr. Walker when the specimens were sent to him for verification (Walker & Ricker 1938). But stonefly collecting was now coming to the forefront and one such outstanding collecting trip was a long transect traverse over the crest of the Cascades, bivouacking without sleeping gear along the route, which began in Canada, topped out at 1800 m, considerably above tree line at the international boundary (Monument No. 48), and ended on the Nooksack Valley floor in Washington. Reaching the road located on the latter, he and pal, Harold Baldwin, hitchhiked up the road to the hamlet of the Glacier, bought breakfast on credit(!) and then hitchhiked back down the road to a point where they could walk across the boundary to enter the south end of the Cultus Lake ('Columbia') valley and then hike 13 more kilometres to home! In his pockets were precious vials filled with stoneflies. Just try and walk across the border today, let alone with any mysterious vials in your pocket!

Well, 1934 was the year that stonefly investigations had begun, and Cultus Lake was a critical locale for this work, harbouring many new species that he described in the scientific literature (Ricker 1939). Bill also collected caddis flies in the area but for those he could not identify he sent to a specialist, Mr. L. Milne. Much to his surprise one collected was a new species named *Glyphopsyche* (later *Neophylax*) *rickeri* as related to him in a letter dated 29 June 1934. This re-doubled the caddis fly effort as well, finding 11 other new species, to be later described by Milne!

The other significant event in 1934 was the attraction to his future wife, Marion. The relationship became very close when they bivouacked near tree line, without sleeping gear once again, on an outing in early September. One of the interesting investigations during the 1934 season was the discovery and study of the habitat of a unique species of frog, *Ascaphus truei*, which has a short tail, hitherto unknown in this part of the world. Over the years other discoveries of the frog have been found in several fast-flowing creeks between West Vancouver and Manning Park to the east, roughly 200 – 250 km in extent. The publication on this unusual species (Ricker & Logier 1935) came to the forefront about 60 years later when the frog was discovered in a creek above West Vancouver by school children, who feared it would be locally extirpated if a subdivision developer modified the stream banks according to his landscaping plans. The Engineering Department of the Municipality of West Vancouver read Bill's publication to fathom the habitat requirement of the frog. They forced the developer to modify the plans and have the stream bank left in its natural state. This is a second example of when ecology became 'of age'!

In 1935, work at Cultus Lake had to be concluded into a Ph.D. thesis, but there were big distractions – namely marriage in March to Marion, and an insatiable quest to work on his stonefly collections. On the first there was the matter of buying a ring ($65.00), a new suit for the occasion ($22.50), presents for the bridegroom and best man ($25.00), and refreshments for the stag ($4.00). It has always been said that the cost of a new suit is about the same as the value of an ounce of gold. Franklin Roosevelt had just pegged the value of the latter at $35.00! For the stoneflies Bill was busy tracking the literature on all of the para – and holotypes of the described North American species, many of which were housed in European institutions. After completing his Ph.D. he was to be on his way to have a look at them as well as to collaborate with the specialists who lived there. Consequently his diary has a hiatus again. There are no entries for the period of 1 January to 23 May 1935.

For the late spring and early summer period of 1935 at Cultus Lake the diary reveals several trips to the local mountains with some botanical observations. On 9 August he and Marion loaded their auto (a Ford Model 'A' roadster), bound for the University of Toronto to wind up the dissertation. They 'botanized' their way to Revelstoke using a back road route through the Douglas – Chapperon Lake ranch country north-east of Merrit and camped 1 night at tree line on Mt. Revelstoke in order to look at the alpine flora in prime bloom time. In the 1930s there was no Trans-Canada highway through or around the Selkirk and Purcell Mountains. The car was loaded onto the passenger train at Revelstoke and disembarked at Golden, British Columbia, on the floor of the Rocky Mountain Trench, before continuing through the Rockies and thence eastward across the prairies to Toronto. Unfortunately, the diary has another hiatus between Golden and the day (16 October 1935) he took his Ph.D. orals at Toronto. Happily the orals were passed although not with the flying colours that Dr. Harkness had anticipated. Following that day the two were on their way to Europe in what could be described as a moving post-doctoral study, based at London and Plön, Germany. It was winter season, not noteworthy botanically, and a stop at Linz, Austria brought forth the only comments on the flora seen throughout the journey.

Bill and Marion returned to Cultus Lake on 23 May 1936 driving from North Bay by way of his favourite route through the USA. The new distraction was the author, riding in a makeshift hammock behind the driver at only a few months old. Over the 12 days of travel the attention was not on natural history stops except to look at geysers in Yellowstone National Park and to swat the bushes for stoneflies. Reaching 'home' the diary yields another 45-day hiatus, the entries finally beginning with his wilderness trips in late summer. There were several including his second multi-day hike into Chilliwack Lake to see the famed hermit, Charlie Lindemann.

In the autumn of 1936 the weather did not perform its customary switch to a cloud-filled valley, oppressed by the Aleutian Low. The pressure centre remained far offshore to the north-west and weather was cloud-free until mid-December. What happened to the biological observations for the year and were there many? The diary is not helpful, mentioning only a few plants, not many more stoneflies, and the customary few big game seen in the American natural parks. Notations in Bill's copy of Henry (1915), however, provides a better insight of the species seen, especially those on the alpine ridges above Cultus Lake.

The year of 1937 was to be his last at Cultus Lake, although there were many subsequent short visits to the area, to about 1997, including two or three with his grandchildren. The lake had several freeze-ups and thaws over the winter. Not much natural history was noted until hiking season began in late May with three trips into the hills and onto the alpine ridges before mid-July, followed by a 2 week family holiday to Mt. Rainier and the desert country to the east of it, returning along the coastline of Oregon and Washington. One stop on the trip was to see Dr. R.B. Lawrence, a renowned dendrochronologist who had just completed a study on the dating of a landslide into the Columbia River. Bill was interested to see if other applications of the technique could be used for his work. They became life-long friends, exchanging ideas on climate change and its biological consequences whenever they corresponded. Botanically, the trip noted many not-seen-before flowers at Mt. Rainier, including *Smelowskia calycina*, a Cruciferae, and *Zerophyllum tenax*, a bear-grass of the Liliaceae. Back at Cultus Lake in August the wild cucumber, *Marah oreganos*, was another new discovery. The daily log of plant and stonefly observations for 1937, however, is not totally revealing of what was taking place over the year, as only 30 species are noted, most of them at Mt. Rainier in the case of flowers.

Inside the diary for 1937, however, is an elaborate multi-page table which shows the plant species seen around Cultus Lake, primarily in that year. For each species the following timeline data is recorded if known:

- Leaves: date – half grown, fully developed, falling off, and gone.
- Flowers: date – first seen, full bloom stage, last seen.
- Fruit: date – first green developed, first ripe, fully ripe, last seen.

As well, there is a space for additional comments and the plant's local name. Bill listed 175 plants which include the upper cryptogams, gymnosperms, monocots, and dicotyledons, as well as one fungus, *Peziza coccinea*, which emerged on 14 March 1937. The table, to put it succinctly, is mind-boggling; the diaries do not hint at such an effort shown by the data compiled for 105 species listed in the table. The 70 species without data are perhaps those seen in previous years, as marked with a date of observation in Henry's (1915) identification guide. Moreover, 175 species is also light; I have found 264 species, excluding the lower cryptogams, by reviewing all of his written records (Table 3), or 313 species if the lower cryptogams are included (Table 2). Noteworthy in the Cultus Lake project is the identification of at least 34 species of aquatic algae, many being planktonic forms. This work on micro-cellular forms was a new break-through in his life as a botanist, but apparently never repeated again.

The Cultus Lake project was steered into the politics of salmon in 1938. Work on the project was to be seconded by a new bi-national organization, the International Pacific Salmon Commission (IPSC), which was to find a way to better manage the depleting stocks in the Fraser River. Bill and his mentor, Dr. R. Earle Foerster, joined the IPSC in that year, moving out of Cultus Lake but visiting it in short sessions over a hectic summer of stream surveys and tagging operations on the river's tributaries and main stem respectively. The work allowed no spare time to look at natural features around him, or if he did, there was no time to insert observations into his field books, which were crammed with creek obstruction and spawning bed diagrams, carcass counts, tag recovery data at canneries, etc.

To Indiana (1939 – 1950)

The academic prowess of Bill Ricker and the operations of the IPSC were not compatible, and with relief his various contacts at Toronto and Nanaimo had put in a good word for him when Dean Fernadus Payne at Indiana University pondered which of three Canadians he should hire to replace one deceased staff member (Dr. W. Scott) and another who took an administrative posting at Stanford University (Dr. D.S. Jordan). Bill's aquatic botany abilities may have been an asset in the competition because part of the job was to work on projects with the Indiana State's Department of Conservation, which were fish population studies on very 'weedy' lakes in the northern glaciated portion of the state. Decreasing fish populations in many, coupled with an ever-increasing growth of aquatic flora and other causes of eutrophication, were the focus of research. Notwithstanding, his ornithological abilities also helped; they needed an ornithology instructor as well, and botany instruction of the aquatic species was part of a limnology course he would also be required to teach. Bill arrived at Indiana University (IU) in early March, ahead of the family, who were left at his parents' home in North Bay while he sorted out the new job and found rental housing. In late May, he shifted north to Winona Lake to re-open an old field research lab-cum-training school for aquatic biology students – three big buildings in all, with dorms, labs, offices, workshop and large storage rooms for equipment including boats. The establishment had not been used for several years over the Depression, and there was much inventory and clean-up work to be done. Immediately Bill noticed a difference in the flora of a new biome, as opposed to southern Ontario, but using the old standby, Gray's Manual of Botany, he quickly identified the unknowns that were crucial to his work. He, however, no longer kept a diary for personal use. If there were botanical observations, they were noted directly in Gray's manual, and possibly in field books used to record endless fish data for each lake under investigation. However, inspection through several of these field books (many for Shoe Lake and Spear Lake) did not reveal any such notations, but possibly a few may turn up if every one is carefully scrutinized.

I was fully aware of the operations in Indiana, asking lots of questions on what he found, where he was going (or we, together), and with time I could detect when he could not recognize a species of biota and had to look it up in a book or manual. In those years I understood the logistics of his field movements, but not the science behind them. However, for most summers at Indiana the family would be moved to North Bay to a better natural (and socio-political) atmosphere. During the war years the politico-patriotism of the American regime was too over-powering for his liking, and for part of the summer at least, he moved the family north of the border, except in 1943 when distant travel was banned and in 1944 when even state-wide travel was impossible due to severe fuel rationing.

Table 2. Regional summary of taxa of flora identified by W.E. Ricker (1923–2001).

Continental region or area	Plant division, class-numbers of taxa[a]							Total taxa each area[b]
	Algae	Fungi moulds	Musci, liverworts	Pteridophytes	Gymnosperms	Angiosperms		
						Monocot	Dicot	
Eastern North America	6	6	13	49	16	160	533	783
S. Ontario (1927)[c]	0	0	0	13	8	95	274	390
Interior North America (plains, mtns., basins)	0	1	0	1	8	10	47	69
West coastal zone, North America	35	3	1	16	28	117	797	997
Cultus Lake, BC[c]	34	2	1	11	15	38	212	313
Northern Eurasia and European Alps	1	5	0	4	8	21	106	143

[a]Identification to species level where possible, or to genera if he encountered taxonomic difficulties, eight common name identifications omitted.

[b]Circumpolar and pan-continental species are in each regional group; hence a grand total cannot be generated by adding up data on the vertical axis.

[c]Data for southern Ontario (1927) and for Cultus Lake are included in the regional summaries as well.

As a family man and a botanist one may ask, 'Was he an avid or good gardener?' The answer is 'no' to both. Gardening was left to his wife, Marion, unless she needed help – which was the case in the last 15–20 years of her life. During the war, however, the authorities promoted back yard 'victory' vegetable gardens to assist the war effort directly, indirectly, and for other subtle motives. Bill complied and in 1944 he planted his garden in a vacant lot near the home at Bloomington. Surprisingly, it was successful although there was some insect attack on a few rows of leafy vegetables and the dreaded worms appeared in some of the tomatoes. Vegetable gardens thereafter, however, were his wife's pursuit soon after the family moved to Nanaimo in 1950.

Botanically, what did Indiana offer? Without a doubt the autumn colours of the hardwoods were impressive to all family members. But for Bill two lesser trees were more significant: the persimmon, *Diospyros virginiana*, of the Ebanaceae family, and the papaw or custard apple, *Asimia* sp., of the Anonaceae family – both with sweet fruit and the former especially delicious after a heavy frost. Notes on whatever else he found are few. Only plants not seen beforehand were ticked off in Gray's Manual of Botany, and compiling lengthy lists of flora had come to an end. Fortuitously, however, there are a few other notes shown on some daily weather records he was maintaining at IU. The purpose of the exercise was not totally clear, but his father would provide him with Tor-

Table 3. Numbers of classification groups of flora, by region determined by W.E. Ricker (1923–2001).

Continental region or area	Taxonomic level[a]			
	No. of families	No. of genera	No. of species	No. of varieties
Eastern North America	112	285	758	19
S. Ontario (1927)	76	227	390[b]	13
Interior North America (plains, mtns., basins)	29	60	68	1
West coastal zone, NA	104	435	958	13
Cultus Lake, BC	65	194	265	5
Northern Eurasia and European Alps	60	125	139	0

[a]Excludes lower Cryptogams.

[b]354 species, excluding all cryptogams, submitted to herbarium collections at the University of Toronto.

onto's climate record and the dates that leaves turned colour or dropped off trees, and again when they along with the flower heads on various forbs re-appeared in spring. It appears that they were trying to gauge the lag time between floral events at Toronto to those at Bloomington, Indiana using the meteorological records as one controlling factor. An additional 20–25 species of plants and trees were found in the notes on this query.

For 1946 and 1947 the records of observations are from long auto trips only, to British Columbia and North Bay, respectively. An intermittent diary for 1948 and a rather complete one for 1 June to December 1949 do yield a few more species. The year of 1949 was a period of overdrive in his life – too many projects, widely separated, bolstered by four trips to the southern Appalachians to collect stoneflies. He had visited these mountains a few times before, netting critical species in his taxonomic work. On one such trip, however, he spotted a very attractive azalea, *Rhododendrum* sp., which never did result in a satisfactory species determination. Previously (1948) he had garnered a flaming azalea, *R. calendulaceum*, while passing through New England on a family trip en route to Bloomington from a summer at North Bay.

During the early course of the summer of 1949 Bill was notified by Dr. J.R. Dymond that the Fisheries Research Board of Canada had an interesting proposition for him. It was in a letter received on 16 June just after he had completed a poisoning report on Lake Manitou, which had too many 'coarse' fish, and the state officials were asking to be fully briefed on the contentious project. He mulled the proposition over at his office at IU, then 'bolted' the next day to Winona Lake. He could not make a decision. After the third trip to the Appalachians in late July he was no farther ahead on the decision. Driving again directly to IU and then onto Winona Lake, where he parked the government vehicle, he continued onto North Bay by rail to see his family. A few days later the family was taken to his old haunts around the Mad River and then to Fred Ide's nearby cabin in the hills behind Collingwood. I suspect he was seeking Fred's advice on the job proposal. Returning to the cottage at North Bay he finally wrote to Professor Dymond about the proposed new job. Five days later (26 August 1949) he was on his way to Smoke Lake, Algonquin Park (another old haunt), to discuss it with the professor at his summer cottage. (Final acceptance, however, still had to be worked out with the Board meetings which transpired in late autumn, when Bill called a meeting to discuss the pros and cons with all family members. It was not an easy decision, nor was there 100% unanimity that he should take the offer.) Departure from North Bay in 1949 was to the west, crossing the international boundary at Sault Ste. Marie and then swinging to the Keewenaw Peninsula of northern Michigan to view the spectacular coast of Lake Superior and to visit a colleague at a nearby 'trout station' on the Otter River. Back at IU he was faced with enrolling students using a new punch card system – the forerunner application to facing up to mainframe computers less than 10 years henceforth. Yet another milestone in his life was the release of his translation from Russian of the Baranov papers on population dynamics in his final year at IU; the text was collated into hand-bound copies by his students. It had been a hectic year, ending when wife Marion was rushed to hospital over the Christmas holidays with a very serious abdominal ailment.

To wind up the botanical facets at Indiana, it is not known how many species he identified within the state. For those species never seen before it is at least 51 but the notations in Gray's Manual of Botany indicate 22 more that had also been seen in Canada. Moreover, there are 60 'ticks' against species in Gray's manual where he neglected to write in a locale of observation. I would suggest that many of these were in Indiana as well, but some could have also been in adjacent states. So the species count of new observations in Indiana could be as high as 121, and perhaps 143 species all told were identified within the state as shown mainly by notation in a field guide.

Nanaimo, B.C. with the Fisheries Research Board of Canada (1950–1973) and retirement beyond (1973–2001)

The move began on 3 June 1950, slowly driving with all family members to the west coast through country not seen beforehand, southwesterly through the United States. The botanical highlights on the long trip were the sequoias, *Sequoiadendron giganteum*, and redwoods, *Sequoia sempervirens*, of California, not only because of their immensity but also because there was some shade. The trip was very warm, if not boiler hot throughout, until the Pacific coast was reached. The trip was not recorded in the traditional diary, though Bill wrote out a report after it was over from notes he had kept in the glove compartment of the car. The era of hard cover book diaries had finished in 1949. Henceforth, the trip report was filed after conclusion of a journey, if reported on at all.

A new tactic on any trip, whatever its length, was to record the observation (if new or noteworthy) of any flora directly into the appropriate natural history guide, usually those of the Peterson Field Guide series. For the flora, two regions of interest in western North America, they are shown by notations in the guides: the Pacific coastal zone and the Rocky Mountain or interior continental series. To supplement awareness and identifications Bill bought local park guides wherever he went; there are about 50 on his book shelves in his study den, and any others acquired but lacked a stiff cover or were substandard in format were discarded at the end of the excursion. Nearly all identifications of the Pacific maritime belt were noted in Niehaus & Ripper (1976), a Peterson Field Guide Series. He also transferred all of his old notations in Henry's (1915) treatise to this modern book, noting the potential taxonomic synonymies, but sorting out the confusion, if in Canada, by reference to Scoggan (1978). There are also ticked recognitions of species in the latter, but he seldom wrote in the locality, which in several instances could have been an eastern North American location. Some trips went from the coastal zone to the continental interior. The two or more biomes were separated when reviewing his travelogue. A separate list of the identified interior species was compiled as indicated in Tables 2 and 3.

For the coastal area there were innumerable short trips on Vancouver Island; there were also many short trips into the south-west mainland corner of British Columbia, including its alpine regions; several mid-distance trips went into the state of Washington and northern Oregon but only a few of the longer trips reached southern Oregon and California. A trip to Hawaii in 1961 did not provide a report and there is no comprehensive field guide for the region on his bookshelf. Otherwise there are reports of the longer coastal trips as follows: 1937 (NW USA), 1953 (NW USA), 1955 (NW USA), 1961 (California), 1971 (NW USA), 1977 (Queen Charlotte Is.), 1978 (California), 1980 (NW Washington), and (1983 NW USA). In other years where there were trips, but no report yet uncovered, the observations of botanical significance had been marked in Niehaus & Ripper (1976). On most trips stonefly collection was the priority pursuit; in a few others it was historical exploration, floral appreciation and birding. Later in life when his eyes could no longer focus under the microscope the stonefly collecting fell to the wayside and more attention was spent on checking out the flora. In 1973 Bill retired from the Fisheries Research Board, when it became a research branch of the re-formatted federal Department of Fisheries and Oceans. He kept a near daily office presence at their Nanaimo facilities until year 2000.

From the 70 years of floral observations found in his books and papers I have found 959 species identified for the western coastal region (Table 3). This excludes the lower cryptogams because most of those are microscopic determinations (Table 2) in the laboratory whereas the higher flora were identified without laboratory procedures or reference to collections in a herbarium. It is noted, however, that 265 species were found around Cultus Lake. That is, since 1950 he has added 693 species to his inventory of self-discoveries. In that time period, however, the focus of interest was narrowed; several families of grass, sedge, rush and other aquatic plants were excluded from further scrutiny.

Interior Basin and range and plateau country of mid-continental North America (1939–2000)

Bill never lived anywhere between Indiana and coastal British Columbia, nor did he ever have long, relaxing holidays in this region, with one possible exception, Texas in 1961. He was always on the move when he crossed the continent, initially by car, but more often by rail or air when on government business in the post-1950 era. Nonetheless, over the years observations in his multitraverses across the continent slowly produced a list of plants of interest to him for this region. Most of the species compiled for this summary (Tables 2 and 3) are taken from trip reports: the pertinent logs for the interior portions of trips in the Pacific Northwest region (1937, 1953, 1955, 1971, 1977, 1980, 1983), Texas and adjacent states (1961), transcontinental excursions (1933, 1934, 1935, 1936, 1946, 1950, 1954), Rocky Mountains of Canada (1958, 1970, 1975), and Wells Gray Provincial Park in Central British Columbia (1957). For other trips into these areas written reports have not been found, nor are there many notations in natural history guides covering these areas, though those found in Cormack's (1967) guide to the flora of Alberta are one such exception.

The tally of species new to him for this region is only 69 (Table 2). Plants seen beforehand elsewhere were seldom registered in his reports or identification guides. No single guide or manual covered the mid-continental regions. However, the more useful one is the Peterson series by Craighead et al. (1963), but within its pages his notations of species identified are for only one trip in the Pacific Northwest (1971) which went as far east as the Grand Tetons in Wyoming. Reviewing the compiled species list, it was the trip to Texas in 1961 which supplied the most new species, but unfortunately the excursion was hampered by a lack of a local guide book for the area of greatest interest, the Big Bend National Park area. Surprising omissions of significance on the mid-continental list are in the following families: Caryophyllaceae, Cruciferae, Saxifragiaceae, Violaceae and Caprifoliaceae.

The manner of travel on long trips can explain some of the reasons for the short list. In family excursions stops during the day were five at the most but three was the usual, namely: one to look at a feature of interest, another to lunch at a picnic table, and the final stop to camp for the night. At lunch break, if spare time was available, the stonefly net was unfurled in preference to other activity, and Bill would swat the bushes along a stream course, searching for stoneflies. If the stop was not near a stream course, a hike with some of the sons usually took place, and he would stop to look at a flower only if he did not recognize its identity. So flora that had been seen elsewhere on the continent did not attract special attention which is borne out by the short inventory of species on the list. On these trips the alpine zone was seldom reached unless there was a day to hike to it, and in the basin and range country this seldom happened unless the vehicle went over a high pass, which was not often. So alpine species were often missed on especially the transcontinental ventures when time was limited. Clearly, botanical identifications were not a priority as it was in the 1930s before the advent of 'plecopterism' and arrival of a family of four sons.

Overseas travels (1935 – 1995)

Bill had more than a fair share of international travel during his life span. Initially it was brought about by the importance of his work in fisheries in the assessment of maximum sustainable yields, be it for individual fish species, or for habitat groups or for that matter the whole biomass of the oceanic environment. However, overseas travel began in 1935 on his own initiative and expense. From the journey he amassed a life-long coterie of about 100 scientific contacts, living in Europe, that would serve him well throughout the following 40 – 50 years of his scientific pursuits. Some were aquatic botanists, a few others were ecologists, several were stonefly specialists, but the majority were involved with fisheries matters. The trip was in winter, not conducive to outside floral exploration.

After the move to Nanaimo in 1950, the international travel picked up: 1956 (Rome and Finland), 1960 (Norway and USSR), 1961 and 1965 (Japan), 1966 (Ecuador and Peru), 1967 (USSR), 1968 (Lappland, Sweden), 1969 (6 months in USSR) and after retirement in 1973: New Zealand (1974), Mediterranean sea ports (1976), Caribbean Sea to west coast of Mexico (1977), garden tour to Netherlands and the UK (1979), Kamchatka (1991) and the European Alps (1995). There were also several other trips to South America. Suffice to say, although trip reports for these travels were prepared, and in the case of the USSR in 1969 they run into 100 or more pages in length (Ricker 1970), the notations on botanical observations are usually scant. The confines of the trip (air planes, trains, and buildings) or their off-season itinerary were limiting factors. Lake Baikal in the USSR (1969) is one exception although a floral guide to the region in the English language is not on his bookshelf.

Another exception to the above noted limitations was the stonefly meeting at Abisko, in Lappland of Sweden (1968). The timing was at high season for alpine flowers. Knowing this, Bill bought a mountain flower handbook for Scandinavia (Gjærevoll & Jørgensen 1963) at a book store in Stockholm while en route. This excursion was a treat; he was heading to an area where Linnaeus had laid the foundations for the binomial classification of organisms. The guide book notes that the mountains of Scandinavia have about 200 species of flora only; the book illustrates 150 and Bill saw 75 species plus another two which were not in the handbook, and hence not identified to species level. Interestingly, of the 75 identified, 61 had been described by Linnaeus, and 13 of those were given a species name of *alpina*, *alpinus* or *alpinum*. Certainly the Abisko trip was a botanical highlight in his career, much more so than that of the guided 'garden tour' in 1979 to Europe. It proved to be of boring visits to rather manicured commercial gardens, finding species that had been imported from British Columbia in some cases! The tally of Eurasian species noted for all trips is 145 (Table 2), and again the majority are dicots, but of a broad array all told as shown by the 60 families they represent (Table 3).

Discussion

A botanical career – how was it side-tracked?

The botanical awareness of W.E. Ricker, or Bill to all (even to his sons), was not well-known among his western colleagues, although their counterparts in the east were fully aware of his prowess in this field. While the fundamentals of the science were acquired as a high school student with additional help from his older sister, it was at the University of Toronto where he excelled in this field. As an undergraduate he enrolled in 13 terms of botany courses, and 13 term courses of biology and zoology; but of the latter, two courses (4 terms) were Genetics and Hydrobiology which obviously also bridged the field of botany. That is, as an undergrad his course work was first to become a botanist and second to be a zoologist, and the fisheries aspect of the latter was scarcely visible in the courses taken. What's more, courses in ornithology and ichthyology were not offered at that time, but at Indiana University he went on to teach the former to undergrads and to provide aquatic botany and fish biology in the limnology course for graduate students.

So the question is asked: 'What happened to an aspiring career in botany?' Certainly, he did not arrive at university to become one at the outset. The course work in the first 2 years was pointing to physics or a derivative thereof: astrophysics, geophysics, or possibly astronomy, and perhaps geology. But physics was the attraction because Dr. Sattersly with the charisma was his favourite professor in those first 2 years. The diaries constantly show this fondness. Nonetheless, Bill 'nailed' the first Christmas exam in botany, and Dr. R.B. Thompson of the Department of Botany was impressed, promising to help him find a summer job. The job which did not materialize was perhaps the beginning of change of direction, because an eager Dr. W.H.K. Harkness in the rival Department of Biology offered him one, as second term time was running out, to a field position which had no bearing on the course he was instructing. However, Bill collected botanical specimens throughout that summer employment period while with Dr. Harkness. By this time Dr. Harkness probably realized that in order to keep Bill on the fisheries track, for which there were hardly any relevant courses, he had to out-manoeuvre the staff in the Botany Department. Bill was already into advanced levels of botany courses in his second year, which could have been a worrisome indicator to Dr. Harkness. He probably realized the need to keep Bill occupied on fishery investigations. So in Bill's second year on campus there was the added incentive of part-time employment offered by Dr. Harkness while Bill was doing a first class performance in Dr. Thompson's comprehensive Spermatophyte course. At the conclusion of the second year Dr. Thompson failed to take the initiative to find Bill a summer job, perhaps knowing that Dr. Harkness already had him set up for trout pond studies, but he planned to put Bill to work as a laboratory demonstrator in the third year.

Again, the astute Dr. Harkness acted. With a summer of countless trout stomachs to analyze in the lab, Bill was the obvious one to do it, if fellow entomology student, Fred Ide, would not undertake the job. That is, while Fred may have been the obvious one to do the task, there was no worry about losing him to the botanist because he was not taking their courses. Dr. Harkness asked Bill to do the work because he possibly feared the worst if Bill became a lab demonstrator for the botanists. However, in third year Bill was not deterred from taking two more 'high-powered' botany courses while struggling with Embryology and lab consuming hours in Comparative Anatomy. Judging by those courses a direction to a career in zoology was still absent at that stage. Wily Dr. Harkness used his bountiful summer student employment opportunities as an inducement to again hire Bill at the end of the third academic year, with work at the new broad-based and well-funded Frank's Bay project. Dr. Thompson did not have a responding 'trump card' to counter with. For fourth year the botany courses were even better, being directly applicable to Bill's aquatic work. But the Department of Biology countered by having Bill demonstrate in *three* labs of Comparative Anatomy (9 hours per week), which completely swept up any free Monday to Friday time, and thereby eliminating any work proposals by the Botany Department. Belatedly, they could only come up with a second term Lab Demonstration position, working with lacklustre engineers, on worse yet, Saturday mornings! Their countermove was too late, too little, and certainly substandard, but he took the job anyway, running his extra-curricular work to 12 laboratory hours of facing students So, Dr. Harkness and his associates won the battle of 'capturing' Bill for graduate work in fisheries biology. The summer work made it an easy choice for him at the end of four years of undergraduate course work, despite the fact that he had better academic credentials to work through the Botany Department.

While Bill was working on the M.A. thesis the grand scheme of Dr. Harknes's manoeuvres began to unravel. Bill was not a successful Rhodes Scholarship contender, much to the professor's dismay. Dr. Walker then entered the 'sweepstakes', offering Bill a career in entomology but an unknown factor kept him from replying. The alternative proposal to have Bill work on a Ph.D. project in New York State unravelled next, much to the chagrin of Dr. Harkness, and now he was really scrambling to find a new direction for his prized grad student. This was the perfect opportunity for the botanists to step in with a counter proposal. Why they did not do so, it will never be known, but perhaps having Roy Cain was enough for their purposes. By a stroke of luck Dr. Harkness with the assistance of J.R. Dymond found a position for him with the Biological Board of Canada at Cultus Lake, as a last gasp direly needed alternative, and from then on is now history. However, the years of outstanding course work of botany did not fall to the wayside, because it was used extensively for the Cultus Lake project where all biological aspects of the study had to be covered. But going to Cultus Lake brought a budding lifetime career in botany to an end. The botany became a tool for other scientific purposes as well as a lifetime hobby.

The question has been asked before: 'How did Dr. W.E. Ricker become such a versatile scientist?' The late Peter A. Larkin responded to such on one occasion by telling the questioner that his broad based university education was the key, as offered in the 1920s but no longer offered then (i.e. 1990s). Bill Ricker had acquired the basic concepts to go any direction in science. When he passed away in 2001 the local Nanaimo press quizzed Dr. Richard Beamish at the Pacific Biological Station as to why his death was so significant to the international scientific world and its media? They locally were in the 'dark'. Dr. Beamish responded: 'He was the best fisheries biologist ever produced in Canada. There was nothing that he could not do' (when asked to do so). Certainly this trait was shown in his botanical prowess throughout his life as well.

Summary – life's end

W.E. Ricker's botanical career began as a teenager, identifying flora in two ecoregions within Ontario. By the time he finished his 5 years at high school (age 17) he had collected and pressed 97 reference species and had identified about that many more. Collections and identifications were made at North Bay in spring and late summer, on the south side of the Boreal Forest, located on Precambrian Shield rocks, favouring acidiphilic species. In early and mid-summer the family moved to a grandparents' farm at Plattsville, Oxford County, near Woodstock in southern Ontario, in an area of farmland, with mixed native hardwood forests, underlain by Paleozoic platform carbonate rocks favouring calciphilic flora. Thus diverse floral assemblages were sampled in those early years. He was already on the learning curve of recognizing ecological adaptation.

Going forth to the University of Toronto, Bill enrolled as an already marked student, having just achieved the highest marks in their compulsory entrance exams. Ironically his highest rankings were in physics and chemistry and thus the first year courses taken were weighed in that direction. Adding to the irony, two terms of botany were taken from the Department of Botany in the first year, as opposed to only one term of biology provided by the zoological dominated Biology Department. Bill elected to accept summer employment from the zoologists (W.K. Harkness) because H.B. Thompson of the Botany Department failed to find him a position as he had intended to do. Nonetheless, while working at Lake Simcoe and at the mouth of the Credit River estuary on Lake Ontario, Bill collected flora for his second year course in botany, amassing 267 species to be combined with those of the previous summer. A reward was to be given to the student who submitted the most correctly identified plants. Bill's final submission was 354 species of spermatophytes, substantially less than the 600 or more submitted by a life-long student friend, Roy Cain.

In the second year at university the selected courses were again weighted to chemistry and physics, two courses in each and there was one course each in botany and zoology. At this stage Bill was acquiring some 'tools' needed for his eventual fate as a fisheries biologist, namely recognition of flora and invertebrates, both of which are critical to the fishes habitat and diet requirements. Academically, he continued to maintain first class standings despite considerable on and off campus distractions during most evenings. This is the hallmark of brilliant student abilities that professors strive to find: they work hard in the classroom with 100% attention, but study little when leaving it, relying on ability to quickly digest and retain the knowledge on the subject gained in the classroom (in short: works hard – plays hard). The summer employment reinforced these abilities. Work on the private trout ponds in southern Ontario required an acute ability to recognize the diverse flora, fauna and physical characteristics of each operation, made doubly difficult when analyzing the contents of fish stomachs. During that busy summer 121 species of flora were identified, many that he had seen in previous summers, but most of the 138 species of aquatic invertebrates identified were new to his repertoire.

Third year courses at university were re-directed towards the biological sciences with botany being a very significant component, Lower Cryptogams and Plant Physiology and an outstanding class essay was prepared for the latter course. The finances for the year were assisted by a scholarship granted by the Angler's Association. For the following summer a multi-disciplinary project, at Frank's Bay, Lake Nippissing and situated on the Canadian Shield, further extended his biological prowness. All aspects of the riparian and aquatic habitats were under inventory, which included identification and collection of requisite specimens; the list of flora enumerated was 166 species, many of which he had not seen before, bringing his total list of identified Ontario species to more than 600.

The final undergraduate year at the University of Toronto was very busy for Bill. Three courses each in botany and zoology, another course in the History of Biology, added to by 12 hours of lab demonstrator work were on the weekly schedule. The botany courses, particularly, were to become life-long tools for many pursuits, including the field work for both post-graduate degrees. At the conclusion of the academic year Bill and J. Tuzo Wilson of later plate tectonic fame vied for the top science awards, two to each.

Field work for a Master's degree, located at Mad River to the northwest of Toronto, took place over the summer of 1930. The work continued over the following autumn and winter sessions at the university, which saw only a minimum of class courses taken. The thesis was a comprehensive study of the ecological aspects of a reach of river located upstream of the Niagara Escarpment, and it was completed in the early spring. During the course of field work 312 floral species were identified and about 250 species of aquatic and riparian invertebrates were enumerated as critical and potential sources of food, both direct and indirect, for the prime species of evaluation, the 'speckled' (brook) trout, *Salvelinus fontinalis*.

Following completion of the Master's thesis, Bill's future plans were in disarray for the summer of 1931. He continued to work for the Ontario Fisheries Research Laboratory while W.K. Harkness tried to quickly find a viable avenue which would keep Bill under his reins as a graduate student. The Rhodes Scholarship bid had failed, and employment in new York was stalled by zealous immigration authorities. Dr. Walker, on the other hand, had hoped he would accept a research proposal in entomology, while the botanists apparently sat by without a concrete proposal of their own. Visits to Frank's Bay in July and north to Lake Temagami and beyond were the bigger projects of the summer and the latter added many new species to his list of recognized plant species for the province. It also included the usual fish surveys, collection of aquatic insects, small mammal trapping and a look at forest succession. In September, while there was yet no concrete offer for his future likelihood, he hitch-hiked to Point Pelee and met a rude awakening to the many plant species he had not seen before, wishing he had brought Gray's Manual of Botany to help him out. Point Pelee is in the Carolinian ecoregion, whereas Plattsville was on the south boundary of the Great Lakes ecoregion, lying to the north.

Returning to Toronto the much awaited letter of an offer to go west to work for the Biological Board of Canada at Cultus Lake had arrived. At the closing out of a day of indecision the head of the Department of Biology, Dr. Bensley, advised Bill to take the job. He was westward bound in mid-October to new biological and geological surroundings, and the unaccustomed autumn weather of West Coast wetness. However, lonely life at Cultus Lake was countered by active study of all aspects of its surroundings which stopped only on the mountain tops, while it descended onto the flood plain of the Fraser River nearby. Limnological work on the lake saw his first serious attempts at quantitative collection and identification of planktonic algae, and the diverse array of invertebrates in the water column. Collection of water bottom and stream side biota also took place; all of it had to be identified. About 60 species of aquatic vegetation were identified, and over the four years the total floral count reached 265 species. It was in the summer of 1934 at Cultus when the collection and identification of stoneflies began in earnest. This pursuit was to curtail his time spent on examination of flora for years to come [until microscope work was no longer possible with his failing close-up eyesight in the 1990s]. Bill finished his thesis work at Cultus Lake in the summer of 1935, returning there in 1936 after a sabbatical in Europe.

In 1937, Bill set out to tabulate the dates of development of critical stages in a seasonal growth cycle of the plants and trees at Cultus Lake, and by the end of 1937 his list of plants, shrubs and trees for the region had reached 312 species, representing 74 families of flora. Work throughout the Fraser Basin in 1938 should have added many more, but the pressure of the new job with the International Pacific Salmon Commission did not provide much free time.

When Bill moved to Indiana in 1939 the compilation of lists of identified local flora came to an end. He reduced the botanical effort to mainly noting species that he either had not seen before, or in about 20 cases to species that were rarely seen elsewhere. So for the 11 years in that state about 71 species had been earmarked on the margin of relevant pages in Gray's Manual of Botany. Those pages reveal that the year 1939 was the most active period for the examination of new flora, although another 60 marked species without notation of locale may add to the flora identified in Indiana.

In 1950 Bill and family moved back to British Columbia, to live at Nanaimo on Vancouver Island. Using Henry's (1915) manual in the first 15 years or so of his return, he noted new discoveries not seen at Cultus Lake on the page margins. Many of these were plants living in the supra-tidal and marine foreshore zones. If the travel was of several days' duration a trip report was sometimes written, in which there were customary notes on the unusual species of flora seen. The local trips in British Columbia were to the mountains and west coast of Vancouver Island, Queen Charlotte Islands, Fraser Canyon, North Thompson valley, and east to the Rockies.

Bill did much travelling out of Canada and his favourite area was to the mountains and coastline of the U.S. northwest, and over the time span of 1950 – 1990 there were many auto trips extending down into northern California, and two or three going much farther south to near Mexico. Combining all observations from 1931 to 1999 for the coastal zones of the three states, plus western British Columbia, about 997 species were identified for the northeast Pacific region. Some trips penetrated the continental interior. The most productive was an automobile trip to Texas in 1961, but it was hampered by a lack of plant guide while travelling to the southern point of the state. Cross-country auto trips over the period of 1933–1954 added only a few new species on each, because stonefly collecting had replaced the stops to study the flora, and he no longer noted flora that was seen beforehand in either eastern or western North America.

Bill travelled overseas on many occasions, beginning with his post-doctoral sabbatical to Europe in 1935 which was to be the destination for many trips to as recently as 1995. Botanical notations were infrequent on such trips with the exception of an excursion to Abisko, Lappland of Sweden in 1968, which saw an easy identification of 75 species at prime floral season. Trips to other continents yielded a few observations from each, except Japan. A plant list for New Zealand has yet to be located but it is almost certain that the trip was floral-oriented because his sisters went with him, and both were keenly interested in the photographic aspects of botany. Accounts of his travels in Ecuador and Peru note a few groups of vegetation but exacting species recognition was not carried out.

All told, the number of floral species identified by Bill over his lifetime likely exceeds 1500. The level of botanical study was intense in his undergraduate years, when courses were taken and surveys were made to carry out an ecological appraisal for each region visited. However, there were no descriptions of new-to-science species although he suspected that some specimens collected could be candidates, because they did not fit published descriptions. Hence, the publication record on botanical taxonomic matters does not exist. One wonders if the Graminae, *Oryzopsis racemosa* (Sm.) Ricker, or mountain rice, was described by Bill? No, it was likely by an older distant relative living in the U.S. whose family roots tie in with the Ricker migration out of Schwabishland to eastern North America in 1830 to escape conscription in the Prussian army. Furthermore, identification of grasses and sedges were often avoided in his floral surveys. So, there are no scientific publications by Bill on botany, although some elements of the science are noted in his early papers on trout diet and habitats.

Bill, however, was a fully-qualified botanist at the academic level; failure to take up the science as a profession can be traced to an energetic zoology professor outmanoeuvring a botany professor throughout Bill's undergraduate years at the University of Toronto. The level of course work in each science was swayed slightly to botany, but his lab demonstration work was mainly for the zoologists who also arranged summer employment for him. In fact, the academic courses were better career-oriented to botany; the university at the time really did not have a fisheries biology or population dynamics course and there was certainly no course in ornithology at which he also excelled.

As life wound down in the 1990s, however, Bill's attention returned to botany. The stonefly work had ceased in the early 1990s when his eyesight rebelled at peering through a microscope, and the ornithology was hampered as his hearing began to fade in the 1990s. With his wife's ashes put to rest on the top of nearby Mt. Benson he assumed the roll as a reluctant household gardener, though actually relishing the autumn day when planting iris with his granddaughter. And in the final weeks before life's end he took great pride at finding yet another salt tolerant plant on a nearby beach that he had not seen before. After life's end, 8 September 2001, Bill's ashes were placed at two of his favourite environments. On the summit of Mt. Benson (elevation 1023 m) behind Nanaimo, his family and friends placed simple wreaths of mountain top plants over his ashes on 10 November 2002, salal and other subalpine shrubbery being the main constituents. At Jack's Point a finger of sandstone which juts into the outer harbour of Nanaimo, more of his ashes were scattered in June 2003 about a park bench just installed to commemorate his years of life as a scientist at Nanaimo. Planted among those ashes were some of the irises he grew in his garden at home, and one of his favourite shrubs, the Ericaceae, *Arctostaphylos uva-ursi*, or the kinnickinick which grows on the summit of Mt. Benson. Local native names were a fascination to this man of exceptional broad interests and prowess, be it natural science or indigenous history.

Bill was a botanist from the outset; that it did not become an employable career on its own was a quirk in academic roulette!

Acknowledgements

Those who encouraged W.E. Ricker's interest in botany are to be congratulated: an unknown high school teacher, his older sister Evelyn Ricker, colleague Roy Cain, and Professors Thompson, Sifton and Jackson of the Department of Botany, University of Toronto. A big assist in the preparation of this report was provided by the following: Darlene and Albert Nickull of Nanoose Bay, B.C.; Christin and Ursula Bux of Accutype Secretarial Services, West Vancouver; the executors of the estate of W.E. Ricker; and David Noakes. Your help is appreciated.

References

Cormack, R.G.H. 1967. Wild flowers of Alberta. Alberta Department of Industry and Development, Edmonton. 99 pp.

Craighead, J.J., F.C. Craighead, R.J. Davis, G.O. Hagen & J.C. Craighead. 1963. A Field Guide to Rocky Mountain Wildflowers: Northern Arizona and New Mexico to British Columbia. Houghton Mifflin, Boston. 277 pp.

Doyle, A.C. 1895. The Parasite, A Story. Harper & Brothers, New York. 143 pp.

Eastham, J.W. 1947. Supplement to the Flora of Southern British Columbia. British Columbia Provincial Museum Special Publication Number 1. British Columbia Provincial Museum, Victoria. 119 pp.

Gjaerevoll, O. & R. Jorgensen. 1963. Mountain Flowers of Scandinavia. Trondheim. 160 pp.

Henry, J.K. 1915. Flora of Southern British Columbia and Vancouver Island: with Many References to Alaska and Northern Species. W.J. Gage, Toronto. 363 pp.

Morton, B.R. & R.G. Lewis. 1917. *Native trees of Canada.* Canada Department of the Interior, Forestry Bulletin 61, 233 pp.

Niehaus, T.F. & C.L. Ripper. 1976. A Field Guide to Pacific States Wildflowers: Field Marks of Species Found in Washington, Oregon, California, and Adjacent Areas: a Visual Approach Arranged by Color, Form, and Detail. Houghton Mifflin, Boston. 432 pp.

Ricker, W.E. 1930. Feeding habits of speckled trout in Ontario waters. Trans. Am. Fish. Soc. 60: 64–72.

Ricker, W.E. 1931. An ecological study of the Mad River and notes on other Ontario streams with special reference to the speckled trout. M.A. thesis. University of Toronto. 88 pp.

Ricker, W.E. 1932a. Physiological changes and the origins of species. Can. Field-Natural. 46: 30–31.

Ricker W.E. 1932b. Studies of speckled trout (*Salvelinus fontinalis*) in Ontario University of Toronto Studies, Biological Series No. 36. Publications of the Ontario Fisheries Research Laboratory 44: 69 – 110.

Ricker W.E. 1934. An ecological classification of certain Ontario streams. University of Toronto Studies, Biological Series Number 37: 1 – 114. Publications of the Ontario Fisheries Research Laboratory 49: 1 – 114.

Ricker W.E. 1935. Statistical treatment of sampling processes useful in the enumeration of plankton organisms. Fisheries Research Board of Canada, Manuscript Reports of the Biological Stations, No. 255. 19 pp.

Ricker, W.E. 1939. A preliminary list of stoneflies (Plecoptera) from the vicinity of Cultus Lake, British Columbia. Proc. Entomol. Soc. Brit. Colum. 35: 19–23.

Ricker, W.E. 1970. Some fishery research stations in the USSR, 1969. Fisheries Research Board of Canada, Manuscript Reports of the Biological Stations, No. 1091. 109 pp.

Ricker, W.E. & E.B.S. Logier. 1935. Notes on the occurrence of the ribbed toad (*Ascaphus truei* Stejneger) in Canada. Copeia 1935: 46.

Robinson, B.L. & M.L. Fernald. 1908. Gray's New Manual of Botany, 7th edition. American Book Company, New York. 926 pp.

Scoggan, H.J. 1978. *The Flora of Canada.* National Museums of Canada, Publications in Botany, No. 7, Vols. 1 – 4. 1711 pp.

Walker, E.M. & W.E. Ricker. 1938. Notes on Odonata from the vicinity of Cultus Lake, B.C. Can. Entomol. 70: 144–151.

Environmental Biology of Fishes (2006) 75: 73–93
DOI 10.1007/s10641-005-2443-x

Bill Ricker's records as an ornithologist

Karl Ricker
3253 Archibald Way, V0N 1B3 , Whistler, British Columbia, Canada (e-mail: accutypeservices@shaw.ca)

Received 19 February 2004 Accepted 15 May 2005

Key words: birds, life list, ornithology, naturalist

Bill's "life list" of identified bird species

Few people are aware of Bill Ricker's "life list" of birds. Most assumed that he was not a "lister", but certainly in his first decade of birding he kept copious lists, and worried that he could not keep the pace with his mentor, Charles Ramsay. On the other hand his lists from Indiana never left the state, so it would seem, and lists only re-appeared with the conclusion of some special trips thereafter. However, in year 2001 while tidying up his offices, I found a hand-written "life list" of birds in his filing cabinets. Undated, I forgot to ask him when he compiled it when I noted a lack of equatorial oceanic species on the list. I discussed this with him. He promised to make an addendum but fatal illness set in within the few days to follow. The compiled list (Appendix Table A1) was drawn up mainly from memory, and because the species seen on the European alpine trip of 1995 are not on the list, whereas all other significant sightings on previous trips are on it (including Kamchatka in 1992) we can guess that it was prepared in 1993, give or take a few months on either side of that year. My checks of Bill's diaries and older boyhood lists, show that his memory was quite good. His list indicates 640 species whereas my check on his work shows 715, give or take a few which have been lost or added in taxonomic revisions. That is, his recall was almost 89.5%, which is excellent for somebody in his 80s. Appendix Table A1 is the amended list.

The following is a summary of species seen by Bill over his 80 years of birding:

- Seen in Canada (364), not seen elsewhere – 75 species
- Seen in U.S.A./Mexico (359 species), new sightings not seen elsewhere – 92 species
- Seen in Japan (26 species), new sightings not seen elsewhere – 19 species
- Seen in New Zealand (64 species), new sightings not seen elsewhere – 42 species
- Seen in Hawaii, Fiji and Tahiti (30 species), new sightings not seen elsewhere – 17 species
- Seen in Peru/Ecuador and Canal Zone (82 species), new sightings not seen elsewhere – 60 species
- Seen in Eurasia (i.e. N. Asia) (160 species), new sightings not seen elsewhere – 99 species
- Seen on two or more continents or Pacific Island and continents – 334 species

The total Species seen 715 speciesThe total is not at all overwhelming for enthusiastic birders, who often travel on specific missions to see birds to add to their repertoire. These types of birders often have a life list of 2000–4000 species, out of a world species total of about 10,000 known at present. Bill's approach to birding was holistic, and he was not out on a mission to go the "extra mile" to find something to add to a list.

At life's end Bill's ashes were scattered at spots about Nanaimo; one at his request was atop Mt. Benson, which he climbed many times, beginning in 1933, and the other at a family-dedicated park bench at Jack's Point, which provides a great view of this mountain – a fitting farewell to our ornithologist.

Appendix Table A1

Life list of bird species seen by W.E. Ricker

Table A1. Life list of bird species seen by W.E. Ricker.

Group & Species	Canada	U.S.A.	Europe	Elsewhere
Loons				
Common loon	ON, BC, QCI	WA	SF	
Pacific loon	BC	WA		
Arctic loon			DK, SF	
Red-throated loon	ON, BC			
Atinamon				
Perdiz serranch (*Nothopracta pentlandii*)				EP
Grebes				
Western grebe	BC, QCI	OR, WA, CA		
Red-necked grebe	BC, ON	OR, WA, etc		
Horned grebe	BC, ON	OR, WA, FL, etc		
Eared grebe	BC, ON, PR	FL, OR, CA	USSR, DK, SWD	
Pied-billed grebe	BC, ON	IN, FL		
Great crested grebe			USSR, SWD	
Dabchick			USSR	
Chilean grebe (*P. chilensis*)				EP
Little grebe (*P. ruticaulis*)			USSR	JP
Crested grebe				NZ
Albatross-shearwaters				
Northern fulmar			AO, Europe	
Black-footed albatross	BC, QCI			
Giant Petrel				NZ
Pink-footed shearwater	BC, QCI			
Sooty shearwater	QCI	WA		
Cape pigeon				NZ
Storm petrels				
Fork-tailed storm petrel	BC, QCI			
Leach's storm petrel	BC, QCI			
Pelicans, Cormorants, etc.				
Man-o'-War (Frigate) bird		FL		MX, EP
Magnificent frigate bird				HW
Red-tailed tropic bird				Tahiti
White-tailed tropic bird				HW
American white pelican	AL	OR, CA, ID, FL, CO, TX		MX
Brown pelican		CA, FL, TX		
Peruvian pelican				EP
Northern gannet	PQ		NO	
Peruvian booby				EP
Brown booby				MX
Blue-footed booby? or other sp.				MX
Black shag				NZ

Great cormorants	NB, PQ		UK, NO	
Brandt's cormorants	BC	WA, OR		
Double-crested cormorants	BC, ON, NB	IN, FL, OR		
Pelagic cormorants	BC			
Common cormorants				JP
Little shag				NZ
Anhinga		FL, TX		
Pied shag				NZ
Brazilian cormorants				EP
Red-legged cormorants				EP
Guanoy cormorants				EP
Spotted shag				NZ
Wading birds				
Capped heron				Canal Zone
Cuca heron (*Ardea cocoi*)				EP
Great blue heron (morph)	NB, BC, ON, QCI, etc	IN, WA, WY, TX, NM, OR		
White heron (morph)		FL, TX		
Louisiana (tri-coloured) heron		FL, TX		
Little blue heron		GA, FL, IN		
Green (green-backed) heron	ON, BC	OH, IN, FL, UT		
Black-crowned night heron	ON, BC	OH, IN, FL		EP
Yellow-crowned night heron		FL		
"Eurasian" heron (*A. cinera*)			UK, USSR, GY	
Great egret		OR, IN, WA, FL, CA, TX		EP
Snowy egret		FL, GA, OR, CA		EP
Cattle egret		FL	GY	EP
Reddish egret		TX		
Striated heron (*Butorides striatus*)				EP
American bittern	ON, BC	IN		
Least bittern	ON			
"European" bittern (*B. stellaris*)			USSR	
White-faced heron				NZ
Wood stork		FL		
White stork			USSR	
Andean ibis (*Plegodis ridgwayi*)				EP
White-throated ibis				Canal Zone
Glossy ibis		FL, GA, IN, FL		
White ibis		FL		MX?
Blue reef heron				NZ, Tahiti
Roseate spoonbill		TX, FL		
Greater flamingo	ON (Zoo)	FL		
Chilean flamingo				EP

Table A1. Continued

Group & Species	Canada	U.S.A.	Europe	Elsewhere
Swans, Geese & Ducks				
Swans & Geese				
Mute swan	BC, ON	NC	UK, ND, DK	NZ
Tundra (Whistling) swan	ON, BC, NWT			
Tundra swan – Bewicke form			SWD	
Trumpeter swan	BC	WY, MT		
Whooper swan (ID?)			GY	
Black swan				NZ
Canada goose	BC, ON	ID, OR, CA, WY, MD, IN, NC, WA		NZ
Brant	BC, ON?			
Greater white-fronted goose	BC	CA, OR, NC		
Blue goose – form		NC		
Lesser snow goose – form	ON	IN		
Greater snow goose	BC	MD, NC, OR, TX, CO		
Red-breasted goose			USSR	
Andes goose (*Chloephaga melanoptera*)				EP
Bean goose				JP
Paradise duck				NZ
Ducks				
Patopuna o yuesa (*Anas versicolor*)				EP
Mallard	BC, ON	WA, IN, WI, WY, OR	UK, SWD, SZ, USSR	NZ
Black duck	ON	ME, WI		
Grey duck				NZ
Northern pintail	BC, ON	WA, IN, NC, OR		
Blue duck				NZ
Gadwall	BC, ON	OR		
American widgeon	BC, ON	NC		
Eurasian widgeon	BC			
NZ Shoveler				NZ
Northern Shoveler	BC	NC		HW
Blue-winged teal	BC, ON	IN, FL		
Cinnamon teal	BC	WY, WA, OR		
Green-winged teal	BC, ON	KS, CO, FL	Europe, USSR	
Wood duck	BC, ON	IN, FL		
Tufted duck	BC		USSR, SF, GY	
Pochard			USSR	
Redhead	BC, ON, PR	OR		
Canvasback	BC, PR, ON	WI		
Ring-necked duck (RN bluebill)	BC, ON	IN, GA?		
NZ scaup				NZ
Greater scaup (bluebill)	BC, ON	WY, MA		
Lesser scaup (bluebill)	BC, ON	WY, NC, GA, FL	Europe	
"Eurasian" scaup				

Common (American) goldeneye	BC, ON	MA	UK	
Barrow's golden-eye	BC	WY		
Bufflehead	BC, ON	MA		
Harlequin duck	BC			
Common eider	PQ		NO, SF	
Long-tailed duck (Old squaw)	ON, BC	IN	UK	
Black scoter (American scoter)	BC			
White-winged scoter	BC	IN		
Surf scoter	BC		AO	
Hooded merganser	ON, BC	WY, East		
*Red-breasted merganser	ON, BC	MA		
Common merganser (Goosander)	ON, BC	NC	DK, NO	
Smew			USSR	
Ruddy duck	BC	OR		
Shelduck			ND	
Pato cabeja castaña (*Netta erythrophthalma*)				EP
Patorana o pato todon (*Oxyura ferruginea*)				EP
Pato de torrentes (*Merganetta armata*)				EP
Mandarain duck				JP
Diurnal raptors				
Vultures				
Turkey vulture	BC, ON	IN, NV, East, NM, TX, OR		EP
Black vulture	BC, ON	IN, NV, East, TX		EP
Andean condor (*Vultus gryphus*)				EP
Hawks, Eagles, Kites				
Buzzard (*Buteo buteo*)			GY	
Aguilucho grande (*Geranoaetus fuscescens*)				EP
Mississippi kite		PI, TX, NM		
Swallow-tailed kite		FL		
Snail kite		FL		
Black kite (*M. migrans*)			USSR	
Black-eared kite				JP
Northern (American) goshawk	ON, BC	TX		
Cooper's hawk	BC, ON	FL		
Sharp-shinned hawk	ON, BC, NB	FL		
(NZ) Harrier (*Circus approximans*)				NZ
Northern harrier (Marsh hawk)	ON, BC	OR, IN, ID, WY, NV, NM		
Harrier sp. (not the above)			Europe	
Aguilucho cordillerano (*Buteo poecilochrous*)				EP
Rough-legged hawk	ON, BC	WY, IO, NB, IN, WI		
Ferruginous hawk	PR	CO		
Red-tailed hawk	ON, BC, PQ	IN, CO, UT, WY, OR, TX, NV, WI, IL, East, NM, AZ		

Table A1. Continued

Group & Species	Canada	U.S.A.	Europe	Elsewhere
Red-tailed hawk-Harlan form	BC			
Swainson's hawk	PR	PL, OR, CO, NV		
Broad-winged hawk	ON, BC, PQ	PL, CO, UT, NV, IN, East		
Red-shouldered hawk	ON	IN, IL, FL		
Harris hawk		TX		
Gray hawk		NM		
Hawaiian hawk				HW
Golden eagle	PR, BC	PL, CO, WY, CA, OR, NM		
Bald eagle	ON, BC, QCI	WA, ME, IN		
Caracara		FL, TX		EP
Osprey (Fish hawk)	PQ, ON, BC, QCI, NB	OR, WY, IN, NC	SWD, USSR	EP
Falcons				
Gyrfalcon	ON			
Prairie falcon	PR	NV, PL, WY, OR UT		
Peregrine falcon (Duck hawk)	BC, ON	OR, WY		
Merlin (Pigeon hawk)	BC, ON, NB			
American kestrel (Sparrow hawk)	ON, BC, etc, NB	IO, NB WY, OR, NM, UT, NV, East, UT		
Eurasian kestrel			UK, AU	EP
Aplomado falcon		NV (?)		EP
Small falcon 2 spp?			GY, USSR	
Upland game birds				
Plain chachalaca		TX, AZ		MX
Black grouse			Europe	
Blue grouse	BC	WA		
Spruce grouse (Willow gr)	ON, BC	WA		
Franklin grouse (form of above sp.)	BC			
Ruffed grouse	ON, BC	IN, East		
Sharp-tailed grouse	PR			
Greater sage grouse		WA, CO, UT, WY		
Gray partridge	BC, ON		UK	
Partridge (*Perdix perdix*)			UK, Moravia	
Greater prairie chicken		IN		
Ring-necked pheasant	BC, ON	CO, WA, IN, East, OR	UK, SWD	NZ, HW
Golden pheasant (form of above sp?)		CO?	SWD?	
White-tailed ptarmigan	BC, AL	WA		
Willow ptarmigan (grouse)			Europe	
California quail	BC	CA		NZ
Mountain quail		AK		
Northern bobwhite	ON	IN, Virginia, TX		
Chukar	BC, PR	WA		

Coturnix quail (*C. coturnix*)			USSR	
Japanese quail (*Coturnix japonica*)				JP
Wild turkey	ON	SD, East US, IN		
Montezuma quail		AZ		
Gruiformes (Cranes, Rails)				
Guarahuru cordillerano o chinalinda (*Phalcobaenus albogularis*)				EP
American coot	ON, BC, PR	GA, FL, TX, OR, CO		HW
Sandhill crane	BC, ON	OR, CA, IN		
Gallineta comun (*Rallus sanguinolenta*)				EP
Purple gallinule	ON (rare location)	FL		
Moorhen (Common gallinule)	ON	FL	UK, DK	EP
Sora	ON, BC			
King rail	ON			
Virginia rail	ON			
Weka				NZ
Limpkin		FL		
Pukeko				NZ
Corncrake			USSR	
"Eurasian" or "Australian" water coot (*F. atra*)			SZ, UK, USSR, GY	NZ
Shorebirds				
"Eurasian" Golden plover			Europe, SWD	
Spur-winged plover				NZ
Mountain plover		NM, TX		
Pacific golden plover	BC			HW?
American golden plover	BC, ON, NB	Plains	UK	
Black-bellied plover	BC, ON, NB	Plains, CA, GA, OR, FL, TX	UK	
Piping plover	ON, PEI, NB		UK	
Semipalmated plover	ON, BC, NB	IN, CA, WA, GA, etc, OR	UK	
Wilson's plover	ON	GA, FL		
Killdeer	ON, BC, PR, ATL	IN, NV, East, etc, OR		MX
Northern lapwing			USSR, UK, NO, ND	
Andean lapwing (*Prilocelys resplendens*)				EP
Banded dotterell				NZ
American oystercatcher		FL		
Black oystercatcher	BC, QCI	WA		
South Island pied oystercatcher				NZ
NZ black oystercatcher				NZ
"European" oystercatcher (*H. ostralegus*)			UK, ND, NO, USSR	
Black-necked stilt		OR		EP, MX
American avocet		NV, TX, OR		
Pied stilt				NZ
Greater yellowlegs	ON, BC	IN, WA, ME, TX?, CA?		
Lesser yellowlegs	ON, NB	IN, WA, East, TX?, CA?		

Table A1. Continued

Group & Species	Canada	U.S.A.	Europe	Elsewhere
Solitary sandpiper	ON	IN		MX
Redshank			SF	
†Spotted sandpiper (*Actitis macularia*)	ON, BC	IN, WA, FL, etc.		
Western willet	PR	OR, Plains, TX, FL, UT, CA		MX
Curlew (*N. arquata*)			NO	
Long-billed curlew		WA, Plains, OR, UT, TX, CA		
American whimbrel	BC			
Upland sandpiper	ON	IN, CO		
Greenshank			UK	
Marbled godwit	PR	Plains, CA		
Black-tailed godwit			ND	
Bar-tailed (Pacific) godwit			ND	NZ
†Playeromonchado (*Actitis macularis*)				EP
Wandering tattler	BC			HW, Tahiti
Surfbird	BC			
Ruddy turnstone	BC, ON, PQ, NB	FL, TX?	Europe	HW
Black turnstone	BC, QCI	TX?		
Curlew (*N. arguata*)			UK	
Common sandpiper			Europe	
Western sandpiper	BC	WA, GA, FL, OR		
Semipalmated sandpiper	BC	WA, GA, FL		
Least sandpiper	ON			
Sanderling	ON, BC, NB	WA, ME, IN, FL, TX, CA		
Pectoral sandpiper	ON, BC	WA, IN, NY		
Rock sandpiper	BC			
Dunlin (Red-backed S/P)	ON, BC	OR		
White-rumped sandpiper	ON, NWT			
Red knot	ON			
*Purple sandpiper	NB	Indiana!!		
Baird's sandpiper	ON	WA		
Wilson's phalarope	BC	WA, OR, IN, CA, WY		EP
Red-necked phalarope (Northern)	BC, NB	WA		
Ruff			SWD	
American woodcock	ON, NS	IN, etc		
"European" woodcock			UK	
Short-billed dowitcher	ON, BC	TX?		
Long-billed dowitcher	BC	Plains, TX		
Wilson's (Common) snipe	ON, BC		Europe, USSR	
Green sandpiper	ON, BC		USSR	

Jaegars & Skuas				
Parisitic jaegar	ON, BC			
Long-tailed jaegar	ON, BC			
Gulls, Terns & Skimmers				
Gulls				
Southern black-backed gull				NZ, EP
(=Kelp gull =Gaviota dominicana) (*L. dominicanus*)				
Glaucous gull	ON, BC, NWT			
Glaucous-winged gull	BC, QCI	WA		
Western gull	BC	WA?, OR		
Great black-backed gull	ON, NB, PQ	East	UK	
Lesser black-backed gull	ON		UK, SWD, SF, NO	
Herring gull	PQ, ON, BC, ATL, NB	IN, WA, East, etc, MH	UK, SWD, SF, AO	
Thayer's gull	BC			
California gull	BC, PR	CA, WA, OR, Plains		
Ring-billed gull	ON, BC, PQ			
*Mew (Short-billed) gull	BC		UK, USSR, SWD	
Heerman's gull	BC	WA, OR, CA		
Laughing gull		ATL, East, TX		
Franklin's gull	PR	Plains		
Bonaparte's gull	ON, BC	WA, IN, East		
Sabine's gull	NWT			
Black-legged kittiwake	ATL, AO		AO, IR, UK	
Black-headed gull			USSR, UK, AO, SWD, SZ	JP
Little gull			USSR	
Great black-headed gull			USSR, SF	
Black-tailed gull			far east USSR	JP
Gray gull (*L. modestus*)				EP
Black billed gull				NZ
Red billed gull				NZ
Simeon's band-tailed gull (*L. belcheri*)				EP
Gaviota andina o tiulla (*L. serranus*)				EP
Terns				
Least tern	ON	NC	USSR	
Arctic tern	NWT		SWD, SF	
Common tern	ON, BC	East	USSR, SF	
Forster's tern	ON	OR, CA		
Caspian tern	ON	IN, OR, ID, WA, CA	USSR	NZ
Black tern	ON, BC	IN	USSR	
Gaviotin sudamericano (*Sterna hirundinacea*)				EP
Black skimmer		East USA, ATL, TX		
Black-fronted tern				NZ
White-fronted tern				NZ
Royal tern		CA, FL, TX		

Table A1. Continued

Group & Species	Canada	U.S.A.	Europe	Elsewhere
Crested tern				Tahiti, Fiji
Sandwich tern		TX		
Alcids				
Common murre	BC, QCI	OR, WA		
Dovekie	ATL, AO			
Pigeon guillemot	BC, QCI	WA, OR		
Black guillemot	NB		NO, SF	
Razorbill auk	ATL		NO	
Tufted Puffin	BC, QCI	WA		
Rhinoceras auklet	BC, QCI			
Cassin's auklet	BC			
Marbled murrelet	BC			
Ancient murrelet	BC, QCI			
Pigeons & Doves				
NZ pigeon				NZ
Spotted dove				NZ, HW
Casa belita (*Gymnopeta ceciliae*)				EP
White-tipped dove				EP
Wood pigeon			Europe, UK, ND	
Band-tailed pigeon	BC	AZ		
Rock dove (Rock pigeon)	ON, BC, etc	IN, WA	UK, etc, USSR	NZ
White-winged dove		TX, AK, NM, AZ		
Morning dove	ON, BC	TX, AK, NM, East, OR		
Common ground dove		TX, FL, etc, NM		
Barred dove				HW
Inca dove		TX		
Madrugadora o rubiblanca (*Zenaidura auriculata*)				EP
Rufus turtle dove (*S. orientalis*)			Europe, USSR?	JP
Parrots & Allies				
Several parrots seen but		?		EP, HW
no species ID on record except:				
Kea				NZ
Andes parroquet (*Bolborhynchus andicolis*)				EP
Cuckoos & Allies				
Groove-billed ani				EP
Smooth-billed ani		FL		
Yellow-billed cuckoo	ON	IN, GA		MX
Black-billed cuckoo	ON	IN, GA		
"Eurasian" cuckoo (*C. canorus*)			USSR	
Greater Roadrunner		AZ, NM, TX		
Owls				
Eastern screech owl	ON	IN, GA		

Western screech owl	BC			
Great horned owl ("Cat" owl)	ON, BC	WA, IN, TX		
Long-eared owl	ON			
Short-eared owl	BC, ON, PR	IL, MA		
Barn owl	BC	GA, FL		
Snowy owl	BC, ON, NWT	WY, ID		
Barred owl	ON	IN		
Northern hawk owl	ON			
Burrowing owl	PR	Plains, FL, SD		
Northern saw-whet owl	BC, ON			
Northern pygmy owl	BC			
Great gray owl	ON			
Boreal owl	ON			
Goat suckers				
Chuck-Will's-Widow		IN, TX, AK or IL		
Whip-poor-will	ON	IN		
Common poor-will		TX		
Common nighthawk	ON, BC	IN		
Lesser nighthawk		NM, TX(?)		
Chotacabras trinador (*Chordelles acutipennis*)				EP
Swifts				
Vencejo andino (*Apus andecolus*)				EP
Black swift	BC			
Chimney swift	ON	IN		
Vaux's swift	BC	CO		
White-throated swift		TX, NM		
European switft			USSR	
Pallid swift			Rome	
Hummingbirds				
Amazilia costeña (*A. amazilia*)				EP
Ruby throated hummingbird	ON	IN		
Anna's hummingbird	BC			
Rufous hummingbird	BC			
Black-chinned hummingbird		CA		
Blue-throated hummingbird		NM		
Rollers				
Roller			USSR	
Unknown Group				
Picaflor de Cura (*Thanmastura cora*)				EP
Picaflor gigante (*Patagona gigas*)				EP
Kingfishers & Hoopoes				
NZ kingfisher				NZ
European kingfisher (*Alcedo atthis*)			UK, AU	

Table A1. Continued

Group & Species	Canada	U.S.A.	Europe	Elsewhere
Belted kingfisher	ON, BC, QCI, etc	IN, WA, East, etc		
Hoopoe			USSR	
Woodpeckers				
Andean flicker (*Colaptes rupicola*)				EP
Northern flicker-redshafted form	BC	IN, OR, AZ		
Yellowshafted form	ON	IN, East, etc		
Pileated woodpecker	ON, BC	IN, Virginia		
Red-bellied woodpecker		IN, IL, MS, East		
Gila woodpecker		AK, AZ		
Ladder-backed woodpecker		AK		
Red-headed woodpecker	ON	IN, NB		
Acorn woodpecker		CA, AZ		
Lewis' woodpecker	BC	?		
White-headed woodpecker		CA		
Yellow-bellied sapsucker	ON	IN		
Red-breasted sapsucker	BC			
Williamsons's sapsucker		AK, CO, AZ		
Hairy (Harris') woodpecker	ON, BC	IN, CO		
Downy (Batcheldor's) woodpecker	ON, BC	IN, East		
Black-backed woodpecker	ON			
Three-toed woodpecker	ON, BC	CA		
Black woodpecker			AU	
Great spotted woodpecker			Europe	
Middle spotted woodpecker			USSR	
Lesser spotted woodpecker			Europe, SWD	
Green woodpecker			Europe, USSR, AU	
Gilded woodpecker		AZ		
Flycatchers				
Says phoebe		WA, UT, CO		
Eastern phoebe	ON	IN		
Eastern wood peewee	ON	IN		
Olive-sided flycatcher	ON, BC			
Fantail				NZ
Yellow-bellied flycatcher	ON			
Acadian flycatcher	ON	IN		
Willow flycatcher (Traill's flycatcher)	BC			
Alder flycatcher (Traill's flycatcher)	ON			
Least flycatcher (Chebec)	ON, BC	IN		
Hammonds flycatcher	BC			

Pacific slope (Western) flycatcher	BC			
Vermilion flycatcher-rubinus ssp		AK, SW, USA, TX		
Vermilion flycatcher-obscurus ssp				EP
Great crested flycatcher	ON	IN, NC		
Great kiskadee		TX		
Scissor-tailed flycatcher		OK, TX, NM		
Eastern kingbird	ON, etc	IN, IL, MS, AK, OK		
Western kingbird	PR, BC	Plains		
Couch's (or *Cassin's*) kingbird		NM, NB, IL, MISS, KS, UT, CO, NB		
(=Arkansas kingbird in his diary)				
"Common" kingbird =?		NB		
Shrikes & Bulbils				
Loggerhead shrike	ON	WY, UT, CO		
Northern shrike	ON, BC	WY		
Red-backed shrike			USSR, SF	
Bull-headed shrike				JP
Red-vented bulbil				Fiji
Vireos				
Gray vireo		CO		
Blue-headed (Solitary) vireo	ON	IN		
White-eyed vireo	ON	IN, FL		
Bell's vireo		IN		
Yellow-throated vireo	ON			
Red-eyed vireo	ON, BC	IN, WA		
Philadelphia vireo	ON			
Warbling vireo	ON, BC	IN		
Jays, Crows & Allies				
Japanese jay				JP
European alpine jay			SZ	
Blue jay	ON, BC, etc	IN, East, etc		
Steller's jay	BC	WA, OR, etc, AZ, NM		
Scrub jay		CA, OR, NV, FL		
Pinyon jay		NM, CO		
Gray jay – Pacific form	BC			
Gray jay – Taiga form	ON, NB			
Gray jay – Whitehead form		CO		
American (Black-billed) magpie	BC, PR	Plains, WA, OR, ID, NM, OR, WY, CO, NV, NB		
European magpie			USSR, ND, UK, NO	
Blue magpie				JP
Yellow-billed magpie		CA		
White-backed magpie				NZ
Black-backed magpie				NZ
Clark's nutcracker	BC, AB	WA, OR, NM, AZ		
Siberian nutcracker			Lappland (NO, SWD, USSR)	

Table A1. Continued

Group & Species	Canada	U.S.A.	Europe	Elsewhere
American crow	ON, BC, PR	IN, Plains, CO, OR, WY, ID, East		
Northwestern crow	BC, QCI	WA		
Fish crow		ATL?, East		
Mexican jay		AZ, NM		
Hooded crow			USSR, NO, IR, DK, GY, UK, SWD	
Rook			UK, IR, DK, NO, USSR	NZ
Jackdaw			USSR, ND, UK, SWD, IR	
Carrion crow			UK, IR, GY, CZ	
Alpine chough			SZ	
Jungle crow				JP
Hawaiian crow				HW
Eastern cenrian crow				JP
Ravens				
Common raven	BC, ON, PQ	CO, MH, WY, IN, TN, AK, OR, NV, TX, NM, UT, etc	Europe, SWD, USSR	
Chihuahua raven		TX, NM, etc		
Unknown Group(s)				
Elepaio (*Chasiempis sandwichensis*)				HW
Apapane (*Himatione sanguinea*)				HW
'I'Iwi (*Vestiaria coccinea*)				HW
Amakihi (*Loxops virens*)				HW
Larks				
Horned lark (Prairie h. lark)	ON, BC	WY, UT, NV, KS, AZ, WA, NM, CO		
Skylark	BC		UK, ND	HW, NZ
Calandra lark			USSR	
Crested lark			USSR	
Swallows				
Welcome swallow				NZ
Golondrina azuly blanca (*Pygochelidon aianoleuca*)				EP
Golondrina de cuevas (*Petrochelidon fulva*)				EP
Barn swallow	ON, BC	IN, NY, FL	Europe	
Cliff swallow	ON, AB	IN		
Violet-green swallow	BC			
Tree swallow	ON, BC	FL		
Bank swallow	ON, BC	IN		
Northern rough-winged swallow	BC, ON			
Cave swallow	ON (rare location)			
Purple martin	ON, BC	IN, WA, East, etc		
European house martin			UK	
Golondrina plomiza (*Orochelidon murina*)				EP
Golondrina negra (*Progne modesta*)				EP

Chickadees & Allies				
Tomtit				NZ
Tufted titmouse		IN, etc		
Bridled titmouse		AZ		
Bush tit	BC	WA		
Great tit			USSR, UK, DK, ND	JP
Blue tit			UK, DK, GY	
Long-tailed tit			NO	
Coal tit			DK, NO, SWD, GY, AU	
Japanese yellow tit				JP
Black-capped chickadee	ON, BC, NB, NF	WA, IN, East	SWD	
Carolina chickadee		IN, East		
Mountain chickadee	BC, AB	WA, UT		
Boreal (Hudsonian) chickadee	ON			
Chestnut-backed chickadee	BC	WA		
Nuthatches & Creepers				
White-breasted nuthatch	ON, BC	CO, IN		
Red-breasted nuthatch	ON, BC, PQ			
Eurasian nuthatch			USSR, SWD, UK	
Brown creeper	ON, BC	IN		
Tree creeper			USSR, UK	
Pygmy nuthatch		AZ		
Wrens				
Rifleman				NZ
House wren	ON, BC	IN		EP
Winter wren	ON, BC, QCI	TN	GY, UK	
Bewicke's wren	BC	IN		
Carolina wren		IN, East		
Cactus wren		AK, AZ		
Rock wren		AK		
Canyon wren		AK		
Marsh wren (Long-billed wren)	ON, BC			
Sedge wren (Short-billed wren)	ON			
Choqueco wren (*Campylorhynchus fasciatus*)				EP
Old world Warblers, Thrushes & Allies				
South american dipper (*Cinclus leucocephalus*)				EP
European dipper (*C. cinclus*)			AU, Lappland (USSR, NO, SWD)	
American dipper (*C. mexicanus*)	BC			
Pallas' dipper (*C. pallasi*)				JP
Black-cap warbler			UK	
Willow warbler			USSR, SWD	
Grey warbler				NZ
Wood warbler			USSR	
Chiffchaff			UK, IR	
Whinchat			USSR	

Table A1. Continued

Group & Species	Canada	U.S.A.	Europe	Elsewhere
Blue-gray gnat catcher		IN, etc		
Golden-crowned kinglet	ON, BC	IN		
Ruby-crowned kinglet	ON, BC	AZ		
Bluethroat			Lappland (USSR, NO, SWD)	
Wheatear			SZ, UK(?), SWD	
Townsend's solitaire	BC, ON, AB	WA		
Mountain bluebird	BC, AB	US rockies, WY, OR		
Eastern bluebird	ON	IN, FL, GA		
Western bluebird	BC	AZ		
American robin	BC, ON, NB, NF	IN, IL, OR, WA, East, WY, CA, AZ, CO		
Eurasian robin = Redbreast (*Erithacus rubecula*)			UK, USSR, IR	
Fieldfare			Lappland, UK	
Mistle thrush			Europe, IR	
Blue rock thrush				JP
Song thrush (*T. ericetorum*)			Europe, ND	
Varied thrush	BC	WA, MT, etc		
Wood thrush	ON	IN		
Song thrush (*T. philomelos*)				NZ
Hermit thrush	ON, BC	IN		
Whitehead				NZ
Swainson's thrush – olive backed form	BC			
Swainson's thrush – gray backed form	ON			
Veery (Wilson's thrush)	ON, BC	TN		
Thrush nightingale			USSR	
Nightingale			Rome	
Eurasian blackbird (*T. merula*)			UK, USSR, SZ, ND, IR	NZ
Andean thrush (*T. chiguanco*)				EP
Redwing = Red underwing (thrush)			UK, SWD	
Whitethroat			Europe	
Sturnids				
Gray starling				JP
European starling	BC, ON	IN, WA, OR	Europe!, UK, SWD	NZ
Crested myna	BC			NZ
Indian myna				HW
Mimids				
Long-tailed mockingbird (Chirco) (*M. longicaudata*)				EP
Northern mockingbird	ON	IN, East, etc, TX		HW
Gray catbird	ON	IN, NC		
Sage thrasher	BC	WA, NV, etc		
Brown thrasher	ON	IN, NC, etc		
California thrasher		CA		

Wagtails & Pipits				
NZ pipit				NZ
Grey wagtail (*M. cinerea*)			Europe, USSR	JP
White or Pied wagtail (=Water wagtail)			UK, USSR, SWD	
Yellow wagtail			USSR	
American pipit (=Water pipit)	ON, BC, NB, NF	WA		JP
Brown tree(?) pipit (=European)			SZ	
Indian tree pipit (*A. hodgsoni*)				JP
Meadow pipit (*A. pratensis*)			SWD	
Silky & Eurasian Flycathers				
Phainopepla		TX		
Spotted flycatcher (=Gray spotted f.c.)			USSR	
Pied flycatcher			USSR	
Waxwings & Accentors				
Bohemian waxwing	BC	WA		
Cedar waxwing	ON, BC	East		
Dunnock			UK	
Honeyeaters				
Bellbird				NZ
Tui				NZ
Wood warblers				
Black and white warbler	ON	IN, CO		
Prothonotary warbler	ON	IN		
Golden-winged warbler		IN		
Blue-winged warbler		IN		
Tennessee warbler	ON, BC?	IN		
Orange-crowned warbler	BC, ON			
Nashville warbler	ON	IN		
Northern parula warbler	ON	IN		
Yellow warbler	ON, BC	IN		
Magnolia warbler	ON	IN		
Cape May warbler	ON	IN		
Yellow-rumped warbler-myrtle form	ON	IN, WY		
Yellow-rumped warbler-audubon form	BC	AZ		
Townsend's warbler	BC			
Black-throated green warbler	ON	IN		
Black-throated blue warbler	ON	IN		
Black-throated gray warbler	BC	AZ		
Cerulean warbler	ON	IN		
Yellow-throated warbler		IN, DC, etc		
Blackburnian warbler	ON	IN		
Chestnut-sided warbler	ON	IN, TN		
Bay breasted warbler	ON	IN		
Blackpoll warbler	ON	IN		

Table A1. Continued

Group & Species	Canada	U.S.A.	Europe	Elsewhere
Pine warbler	ON			
Prairie warbler		IN		
Yellow palm warbler	ON	IN, FL		
Ovenbird	ON	IN		
Northern waterthrush	ON	FL		
Louisiana waterthrush		IN		
Common yellowthroat	ON, BC	IN		
Yellow-breasted chat		IN		
Kentucky warbler	ON	IN		
MacGillivray's warbler	BC			
Morning warbler	ON	IN		
Connecticut warber		IN		
Hooded warbler		IN		
Wilson's (Black-capped) warbler	ON, BC	IN		
Canada warbler	ON	IN		
Worm-eating warbler		IN		
American redstart	ON	IN		
Silvereyes				
Silvereye				NZ
Tanagers, Cardinals & Allies				
Brazilian cardinal				HW
Saltapolito o chivillo (*Volatina jacarina*)				EP
Western tanager	BC	WA, NM		
Scarlet tanager	ON	IN		
Summer tanager		IN, MS or AK		
Pepitero amarillo grosbeak (*Pheacticus chrysopeplus*)				EP
Cardinal (Northern cardinal)	ON	IN, IL, MISS, KS, East, etc, TX		HW
Blue grosbeak		FL		
Rose-breasted grosbeak	ON			
Black-headed grosbeak	BC			
Dickcissel		IN, IL		
Greenfinch			UK, SWD, DK	NZ
Dominiquiperuana (*Poospiza rubecula*)				EP
Indigo bunting	ON	IN, IL		
Lazuli bunting	BC	SW USA		
Painted bunting		SE USA		
Trile altoandino a chirique cordillerano (*Sicalis uropygiolis*)				EP
"Eurasian" bullfinch			SWD	
Jilguero de cabeza negra (*Spinus ortagellanicus*)				EP
Jilguero negro (*Spinus atratus*)				EP
Japanese yellow bunting				JP
(Siberian) Meadow bunting				JP

Emberizine sparrow & Allies				
Spotted towhee	BC			
Eastern towhee	ON	IN, East		
Green-tailed towhee		OR		
Reed bunting			UK, SWD	
American tree sparrow	ON	IN, WY, East		
Field sparrow	ON	IN, GA		
Clay-coloured sparrow		CO, UT, NV		
Chipping sparrow	ON, BC	IN, East		
Grasshopper sparrow	ON	IN, NV		
Henslow's sparrow	ON	IN		
Savannah sparrow	ON, BC	IN, East		
Vesper sparrow	ON	IN, East		
Lark sparrow		IN		
Harris' sparrow	NWT			
Golden-crowned sparrow	BC, QCI, AB, ON?			
White-throated sparrow	ON, BC	IN, East		
Dark-eyed junco – Oregon form	BC			
Dark-eyed junco – Slate-coloured form	ON, BC	IN, East, etc		
Dark-eyed junco – White-winged form		SD, WY		
Swamp sparrow	ON, NB, BC, NF	IN		
Lincoln's sparrow	ON, BC			
Song sparrow	ON, BC, QCI	IN, East		
Fox sparrow	ON, BC	IN, East		
Lark bunting		OK, TX, CO, NB, WY		
Snow bunting (Snow flake)	ON, BC		SWD	
Lapland longspur	ON			
Smith's longspur	ON			
Pichisanka, gorrion americano, o chingolo (*Zonotrichia capensis*)				EP
Yellowhammer			AU, UK	NZ
Ortolan bunting			USSR	
Icterids				
Western meadowlark	BC, etc	IN, KS, CO, UT, NV, NM		
Eastern meadowlark	ON	IN, MS, OH, East		
Bobolink	ON, BC, QCI?	IN		
Brown-headed cowbird	ON, BC	IN, East		
Yellow-headed blackbird	BC, PR	Plains, NV, WY, OR, CA		
Red-winged blackbird	ON, BC	IN, WA, NV, CO, WY, UT, East, OR		
Rusty blackbird	ON	IN, East		
Brewer's blackbird	BC, PR	Plains, NV, OR		
Common grackle (Purple, Bronzed)	ON	IN, KS, CO, East		
Great-tailed grackle		TX		
Boat-tailed grackle		FL, NC, SC, GA, TX		
Diglossa carbonosa				EP

Table A1. Continued

Group & Species	Canada	U.S.A.	Europe	Elsewhere
Negro o Tordo grande				EP
Orchard oriole		IN		
Scott's oriole		AK		
Baltimore oriole	ON	IN, IL		
Bullock's oriole		AK, NV, AZ		
Golden oriole			USSR	
Finches				
Brambling			SWD	
Chaffinch			DK, GY, SZ, USSR, UK, NO	NZ
Evening grosbeak	ON, BC	AZ, OR		
Pine grosbeak	ON, BC			
Strawberry finch				HW
Gray-crowned rosy finch	BC	WA, WY(?), CO		
Purple finch	ON, BC, QCI?	IN, AZ(?), WY(?)		
Caissin's finch	BC	AZ, TX		
House finch	BC, QCI	AZ(?)		HW
Common redpoll (=Greater r/p)	ON			
Hoary redpoll	ON			
Mealy redpoll			DK, SWD	
Pine siskin (*C. pinus*)	ON, BC	OR		
"Eurasian" siskin (*C. spinus*)			DK	
American goldfinch	ON, BC	IN		
European goldfinch			UK, Rome	NZ
Red crossbill (=American c/b)	ON, BC	AZ, WY, OR		
White-winged crossbill	ON, BC			
Linnet			USSR, UK	
Twite			DK	
Japanese ground linnet				JP
Weaver birds (Old world sparrows)				
House (English) sparrow	ON, BC	IN, WA, East, etc	UK, SWD, USSR, SK, GY	EP, NZ, HW
Eurasian tree sparrow			SWD, DK	JP
Italian sparrow			Rome, SZ?	
Spanish sparrow			Spain	
Unknown Grouping				
Japanese whiteye				HW
Nutmeg mannikin				HW

* Publications by W.E. Ricker on this species of bird.
? Identification is questionable, or the geographic location is questionable.
\# Species count; sub-species, forms, races and variants are not counted (including colour morphs)
etc. = Broad observation distribution; several states or provinces.
Europe: country of species identification not specified.
East: eastern seaboard states of U.S.A.
Total = 715 species + 15 forms + unknown status on 1 or 2.

The 1st Draft was compiled by W.E. Ricker, ca. 1992–1994, amended by Karl Ricker 2003 (revised: March 2004).

The order of listed birds or families of birds follows the arrangement used in the Sibley Guide of North American Birds, National Audubon Society (2000). Old World and species of the southern hemisphere may not be correctly fitted into the arrangement. The life list lacks the records of observations from a Caribbean – tropical Pacific cruise; only four were found in his field guide to the Central American region.

Geographic abbreviations are as follows:

AK – Arkansas, AL – Alberta, ALP – European alpine countries, AO – North Atlantic Ocean, ATL – Atlantic provinces of Canada, AU – Austria, AZ – Arizona, BC – British Columbia (general), CA – California, CB – Caribbean, CO – Colorado, CZ – Czech/Slovak, DC – District of Columbia, DK – Denmark, EP – Ecuador / Peru, EU – Europe, FL – Florida, GA – Georgia, GY – Germany, HW – Hawaii, ID – Idaho, IL – Illinois, IN – Indiana, IO – Iowa, IR – Ireland, JP – Japan, KS – Kansas, MA – Massachusetts, MD – Maryland , ME – Maine, MH – Michigan, MS – Missouri, MT – Montana, MX – Mexico, NC – North Carolina, NB – Nebraska, NB – New Brunswick (Canada), ND – Netherlands, NE – New England, NM – New Mexico, NO – Norway, NV – Nevada, NWT – Northwest Territories and Nunavut, NY – New York, NZ – New Zealand, OH – Ohio, OK – Oklahoma, ON – Ontario, OR – Oregon, PL – Great Plains (several states), PQ – Quebec, PR – Prairie provinces of Canada, QCI – Queen Charlotte Island archipelago, SC – South Carolina, SD – South Dakota, SF – Finland, SWD – Sweden, SZ – Switzerland, TN – Tennessee, TX – Texas, UK – Scotland / England / Wales, USSR – Russia and adjacent countries, UT – Utah, WA – Washington State, WI – Wisconsin, WY – Wyoming. Table 1

Environmental Biology of Fishes (2006) 75:95–110
DOI 10.1007/s10641-005-2444-9

Curiosity, recruitment, and chaos: a tribute to Bill Ricker's inquiring mind

Jon T. Schnute
Fisheries and Oceans Canada, Pacific Biological Station, 3190 Hammond Bay Road, Nanaimo, BC, V9T 6N7, Canada (E-mail: schnutej@pac.dfo-mpo.gc.ca)

Received 17 June 2004 Accepted 27 October 2004

Key words: mathematics, models, biology, fishery management, Ricker curve, yield per recruit, marking experiment, astronomy

Synopsis

Through three versions of a handbook on computations for biological statistics of fish populations, W.E. "Bill" Ricker played a pivotal role in founding the field of quantitative fishery science. His interests, however, extended far beyond the confines of quantifiable events to a deep appreciation for the natural world. In this article, I trace his development of fishery models from the 1940s to the 1970s, using examples that illustrate his approach to statistics and biological systems analysis. I describe changes in technology and statistics that have made it possible to extend his research in new directions, although his approach still lies at the core of all modern fishery models. His gentle, inquiring spirit persisted long after his retirement in 1973, as I illustrate from personal experiences with him during the 1990s.

Introduction

In a historical survey of population dynamics models in fisheries, Quinn (2003, p. 355) concluded that "nobody has made more contributions to the development of fisheries models than William E. Ricker." To support this claim, Quinn cited the breadth, long duration, and statistical focus of Ricker's work. Generations of fishery scientists regularly consulted Ricker's handbooks to find appropriate methods of data analysis. Ricker's (2006) autobiographical sketch in this issue of Environmental Biology of Fishes places his quantitative work in the context of a much broader career. Based partly on my personal experiences with him, I write this companion article as a tribute to his spirit of inquiry, which extended far beyond the confines of quantifiable events to a deep appreciation for the natural world.

The famous handbooks began with an early compilation of numerical methods (Ricker 1948) and grew through two revisions (Ricker 1958, 1975). Only the 1958 edition actually includes the word handbook in its title. Ricker (2006) himself called them the "Green Books," with covers that went sequentially from light green to dark green to aquamarine. He also edited a multi-author handbook that he carried through two editions (Ricker 1968, 1971), working with other outstanding fishery scientists of that time. All these books, combined with extensive journal publications, firmly established his central role in the development of quantitative models for fish populations.

Ricker's (2006) autobiographical sketch reveals another side of this remarkable man. In simple language, he demonstrates his great interest in the natural world around him. He sees organisms in the context of geology, hydrology, and other factors that influence their diversity and evolution. He relates personal experiences with vivid historical details that reflect his passion for discovering the truth. Unresolved questions stay with him. For example, he mentions school days spent in North Bay, Ontario along Lake Nipissing, "whose extensive beaches shoal out into the water with three or four underwater sandbars along the way – a phenomenon for which I have not yet seen the physical explanation." Part of his search entails mathematical techniques for estimating the "vital statistics" of populations, but the driving force comes from his curiosity about nature itself.

The down to earth language of his sketch accurately represents his informal, human style. His friends and acquaintances knew him as "Bill", and I shall often use his familiar name here. He retired as scientist at the Pacific Biological Station in Nanaimo, British Columbia, 3 years before I began working there in 1976. Because he retained an office and used it frequently, I didn't realize at first that he was, in fact, retired. Although I knew of his outstanding reputation, it took me years to appreciate the scope of his achievements. I began working in fisheries as a naive mathematician, with much to learn about biology and the broader world of scientific inquiry. Bill had used mathematics as a scientific tool for understanding nature, based on a wealth of personal experience. He knew well the limitations of models, which can give deceptive estimates and predictions based on assumptions that might be wrong. The Preface to his second Green Book (Ricker 1958, p. 14) contains timeless advice for every fishery scientist:

> "... the practising biologist quickly discovers that the situations he has to tackle tend to be more complex than those in any Handbook, or else the conditions differ from any described to date and demand modifications of existing procedures. It can be taken as a general rule that experiments or observations which seem simple and straightforward will prove to have important complications when analyzed carefully – complications which stem from the complexity and variability of the living organism, and from the changes which take place in it, continuously, from birth to death.

In this article, I trace the development of fishery models through the three handbooks, with a particular focus on statistical methods and biological systems analysis. I discuss Ricker's work in the context of science and technology from the 1940s to the 1970s. Since then, changes in statistical theory and computing have made it possible to extend his research in new directions, but his approach still lies at the core of all modern fishery models. Finally, I relate a few personal experiences with Bill that illustrate his inquiring spirit, always engaging to friends and colleagues. Some examples include mathematical details, often compiled into tables. Equation numbers reflect this style; for instance, (T2.3) refers to equation (3) in Table 2. I think readers can safely skip any mathematical content that proves troublesome.

The Green Books and beyond

Bill's account of his field experiences reminds us of technology very different from that available today. For example, he and Fred Ide explored a region of southern Ontario (Ricker 2006) in a vintage 1922 model-T Ford. He recalls starting it with a crank and driving it up steep hills in reverse to compensate for a gas line fed by gravity from a tank under the front seat. I was 8 years old when Bill published his first Green Book in 1948, and many adults around me could remember driving such vehicles. Some older cars still used cranks for starting, a procedure that could break someone's arm if things went wrong. I can also remember mathematical technology of the period. My parents used a hand-cranked adding machine in their florist business, and it seemed like a big deal to convert to a model operated by an electric motor. I regarded slide rules, especially the thick ones with log–log and trigonometric scales, as the ultimate mathematical hardware.

In 1948 the modern theory of statistics was still relatively young. Fisher published his first textbook on statistical methods in 1925, his genetic theory in 1930, and his treatise on experimental design in 1935 (Fisher 1925, 1930, 1935). Ricker (2006) cites "Fisher's brilliant reconciliation of Mendelian genetics with natural selection" as one of only two major breakthroughs in evolutionary theory since Darwin. Thanks to Bill's generosity (Figure 1), I now have his original copy of Fisher's book on experimental design (2nd edition, Fisher 1937). The inscription suggests that Bill acquired this book in 1941, just a few years before writing his first handbook.

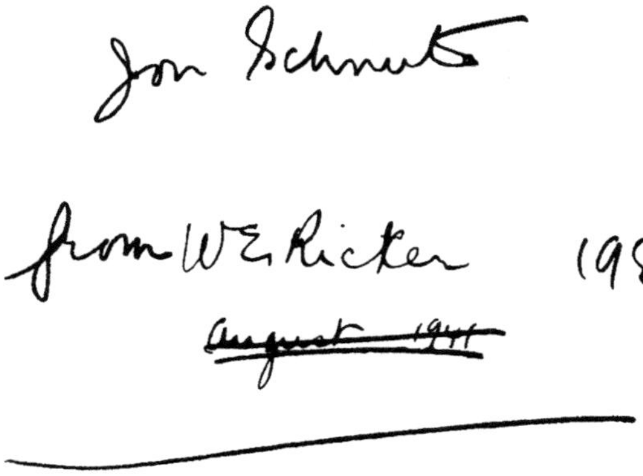

Figure 1. Inscription on the front page of Fisher (1937), when Bill Ricker gave his copy of that book to the author. Bill had originally written "W. E. Ricker, August 1941." Later he scratched out the date and modified the text to read "Jon Schnute from W. E. Ricker 1981." Apparently, Bill acquired the book in 1941 and gave it to the author 40 years later.

These historical details illustrate the world of science and technology in the 1940s, when Bill began writing his Green Books. Ricker (1948) focused on the "vital statistics" of fish populations, including abundance and mortality due to natural and human causes. He also pointed out (*op. cit.*, p. 2) that growth may play a role equal to that of mortality in "a synthesis which leads to conclusions of great theoretical and practical interest." This small volume, however, dealt with abundance and mortality much more than growth. It described key estimation methods, such as measures of abundance based on marking experiments.

The second Green Book (Ricker 1958) introduced a longer list of "vital statistics." In addition to abundance and mortality, these included growth, recruitment, and surplus production. The new topics led naturally to important fishery management issues, including yield per recruit calculations (Chapter 10), relationships between stock and recruitment (Chapter 11), and equilibrium yield for a given fishing rate (Chapter 12). Statistical analyses, largely absent in the first book, appeared frequently in the second. For example, where the former gave formulas estimating abundance, the latter went on to describe methods of estimating uncertainty. I suspect that this progression followed the movement of scientific thinking from the 1940s to the 1950s. First, biologists wanted to know feasible methods for measuring fish populations. Once these concepts and practices became firmly established, the need for assessing uncertainty became more compelling. Many of the statistical techniques reported by Ricker in 1958 had not been invented when he wrote his first book in 1948.

The Green Books show an interesting language transition from "vital" to "biological" statistics. The first uses "vital' in its title, but the second and third use "biological." The opening sentence of the second book (Ricker 1958, p. 17) reads: "The topics which can be considered as vital statistics of a fish population include the following: ..." The third book starts with this same sentence, except that "vital" has been changed to "biological." Ricker's perspective clearly broadened through the years, starting from an initial focus on mortality and abundance tables similar to those compiled for other animal populations, including humans. He gradually extended this framework to encompass more aspects of fish biology, such as growth and recruitment. As predicted in his first book (Ricker 1948, p. 2), adding growth to the mix did indeed lead to "a synthesis ... of great theoretical and practical interest." Perhaps the realization that he was dealing with an entire biological system caused him to replace "vital" with "biological" and to include the word "interpretation" in the title of his third book (Ricker 1975). At last, readers could not only calculate biological statistics, but also interpret them in the context of a larger system.

To give this historical analysis more substance, I follow the progression of two distinct threads through the Green Books. First, Ricker's analysis of a Peterson marking experiment illustrates his approach to statistics. Second, his interest in the complete biological system finds particularly elegant expression in his calculation of fishery yield. In both cases, he compiles the extensive literature of the time into a single reference book for the benefit of his readers.

Peterson marking experiments

Ricker (1948, p. 39) cites Peterson (1896) for the idea of tagging fish to estimate abundance. In the simplest experiment, M marked fish are released into a closed population. Assuming that these distribute randomly and that R of them are recaptured within a total catch C, then

Table 1. Two statistical models for a Peterson marking experiment. A group of M marked fish are randomly dispersed among a population of total size N, and a sample catch of C fish includes R recaptured fish that have been marked. The hypergeometric model accounts for sampling without replacement. The binomial model assumes a fixed recapture probability $p = M/N$, thus ignoring effects due to non-replacement. Ricker (1958, p. 84; 1975, p. 78) cites the results (T1.9)–(T1.10) from the binomial model, which appear as his equations (3.4)–(3.5) in the second and third Green Books. Probability distributions depend on binomial coefficients, defined mathematically in terms of factorial functions: $\binom{a}{b} = \frac{a!}{b!(a-b)!}$

Hypergeometric model

$$P(R|C,M,N) = \binom{M}{R}\binom{N-M}{C-R}\Big/\binom{N}{C} \tag{T1.1}$$

$$\mathrm{E}[R] = \frac{M}{N}C \tag{T1.2}$$

$$\mathrm{V}[R] = \frac{M(N-M)(N-C)}{N^2(N-1)}C \tag{T1.3}$$

$$\hat{N} = \frac{M}{R}C \tag{T1.4}$$

$$\hat{\mathrm{V}}\left[\frac{1}{\hat{N}}\right] = \frac{R(C-R)(M-R)}{M^2C^2(MC-R)} \tag{T1.5}$$

Binomial model

$$P(R|C,M,N) = \binom{C}{R}\left(\frac{M}{N}\right)^R\left(1-\frac{M}{N}\right)^{C-R} \tag{T1.6}$$

$$\mathrm{E}[R] = \frac{M}{N}C \tag{T1.7}$$

$$\mathrm{V}[R] = \frac{M(N-M)}{N^2}C \tag{T1.8}$$

$$\hat{N} = \frac{M}{R}C \tag{T1.9}$$

$$\hat{\mathrm{V}}\left[\frac{1}{\hat{N}}\right] = \frac{R(C-R)}{M^2C^3} \tag{T1.10}$$

$$N = \frac{M}{R}C \tag{1}$$

gives a reasonable estimate of the total population N. For example, if 10% of the tags are recovered ($M/R = 10$), then $N = 10C$ because the catch C must be a 10% sample of the population. In stating this example, I have used notation from later Green Books, for consistency with the discussion below. The symbols (N, M, C, R) here correspond, respectively, to (P, N, H, R) in Ricker (1948).

Although the first book gives only the formula (1), the second includes

- the abundance estimate (1) stated with proper statistical notation $\hat{N}$ (Ricker 1958, p. 84, equation (3.5)), and
- an estimate of the variance $\mathrm{V}[1/\hat{N}]$ (*op. cit.* p. 84, equation (3.4)).

I present these in Table 1 as equations (T1.9)–(T1.10). The handbook cautions the reader that

> "values of MC/R are not very symmetrically distributed, whereas those of R/MC are; hence if the normal curve of error is used to calculate limits of confidence, it is best to calculate them for $1/\hat{N}$... and then invert them in order to obtain limits for $\hat{N}$."

Technically the variance of $\hat{N}$ is infinite, although Ricker (1958) does not mention this fact. Because the number of recoveries R appears in the denominator of (1) and the value $R = 0$ occurs with non-zero probability, the estimate $\hat{N} = \infty$ also has finite probability. Consequently, $\mathrm{V}[\hat{N}] = \mathrm{E}[(\hat{N} - N)^2] = \infty$.

Table 1 gives some perspective to this analysis. The number R of fish recaptured is a random variable, determined by the unknown population N and known numbers M and C of fish marked and caught. Because sampling takes place without replacement, the hypergeometric distribution (T1.1) theoretically represents the distribution of the random variable R. This implies the expected value E[R] and variance V[R] in (T1.2)–(T1.3), known from statistical literature (e.g., Mood and Graybill 1963, pp. 110–113). The estimate (T1.4) for $\hat{N}$ comes from equating the observed value R to its mean value (T1.2). The calculation

$$V\left[\frac{1}{\hat{N}}\right] = V\left[\frac{R}{MC}\right] = \frac{1}{M^2C^2}V[R] = \frac{(N-M)(N-C)}{MCN^2(N-1)} \tag{2}$$

gives an exact formula for the variance of $1/\hat{N}$, and the estimated variance (T1.5) comes from substituting the estimate $\hat{N}$ into (2). This analysis depends on the fact that $1/\hat{N}$ is proportional to the observed random variable R, whose statistical properties are known exactly.

Results presented in Ricker (1958, pp. 83–85) depend on a binomial model for R, in which tags are recovered with constant probability $p = M/N$. This simplifying assumption ignores changes in probability that occur while collecting samples without replacement. The results (T1.7)–(T1.10) follow logically from the model (T1.6), just as (T.2)–(T.5) follow from (T1.1). Both models give identical estimates $\hat{N}$ in (T1.4) and (T1.9). The estimated variance (T1.5) from the hypergeometric model is smaller that the corresponding variance (T1.10) from the binomial model by the factor

$$\frac{1-R/M}{1-R/(CM)} \approx 1 - \frac{R}{M}\left(1-\frac{1}{C}\right).$$

Consequently, Ricker's approximate model predicts somewhat greater uncertainty than the theoretically exact hypergeometric model. This extra margin for error might well be justified, given the logistic difficulties that accompany most field studies. Furthermore, true to his style of comprehensive scholarship, Ricker (*op. cit.*) discusses other approaches, such as the following estimates of N proposed Bailey (1951) and Chapman (1951), respectively:

$$\hat{N}' = \frac{M(C+1)}{R+1} \text{ and } \hat{N}^* = \frac{(M+1)(C+1)}{R+1}. \tag{3}$$

Unlike (1), each of these defines a finite estimate $\hat{N}$ when $R = 0$.

Similar analyses appear in the third handbook (Ricker 1975, pp. 77–81), along with a reference to more recent work by Robson & Regier (1964). Strangely, the hat symbol indicating a statistical estimate has been dropped, so that $\hat{N}$ becomes simply N, as in the first book. I don't know if this occurred by error, or if Ricker felt that his readers would prefer a simpler notation.

Yield calculations

Bill's son Karl kindly gave me a box of Bill's papers and books that seemed relevant to my interests. Among these, I found a paperback (Wilimovsky & Wiklund 1963) with tabulated values of the incomplete beta function $B(x; p, q)$ for

- x ranging from 0 to 1 (in intervals of 0.01),
- p from 0.125 to 35.0 (in varying intervals that increase as p becomes larger),
- q from 3.5 to 4.5 (in intervals of 0.125).

The book has a three-page preface, partly devoted to an explanation of floating point notation (e.g., 8.8E−02 = 0.088), followed by 291 pages with about 115000 tabulated values. The references cite an earlier, more extensive compilation by Pearson (1948) with 494 pages. So why did Bill have this remarkable book, a testament to the realities of computing in 1963?

The answer lies in a particular method of calculating biomass yield from a fishery (Table 2). Consider a single cohort of fish that starts at a specified age a and experiences constant mortality rates F and M due to fishing and natural causes, respectively. As shown in (T2.1)–(T2.2), the total mortality rate $Z = F + M$ determines the population N_t at age $t \geq a$, where abundance declines exponentially from the initial value N_a. Suppose also that fish at age t have weight w_t given by (T2.3), a formula motivated by combining a von Bertalanffy growth model

Table 2. Biomass yield calculation for a cohort that experiences constant rates of natural mortality M and fishing mortality F. These combine to give the total mortality rate Z. Starting at a specified reference age a, the population N_t at age t declines exponentially from its initial size N_a. Fish weight w_t follows a modified von Bertalanffy growth relationship with parameters (W_∞, K, t_0, b), where b accounts for an exponential relationship between length and weight. The fishery captures fish above the recruitment age t_R, where $t_R \geq a \geq t_0$. A formula for the total yield Y depends involves the incomplete beta function $B(x; p, q)$, which reduces to a rational function when q is a positive integer (e.g., $q=4$). Consequently, Y can be expressed analytically in the special case $b=3$, when weight is proportional to length cubed.

Model assumptions

$$Z = F + M \tag{T2.1}$$

$$N_t = N_a e^{-Z(t-a)} \tag{T2.2}$$

$$w_t = W_\infty \left(1 - e^{-K(t-t_0)}\right)^b \tag{T2.3}$$

$$r = t_R - t_0 \tag{T2.4}$$

$$Y = F \int_{t_R}^{\infty} w_t N_t \, \mathrm{d}t \tag{T2.5}$$

Incomplete beta function

$$B(x;\, p,\, q) = \int_0^x u^{p-1}(1-u)^{q-1} \mathrm{d}u \tag{T2.6a}$$

$$= x^p \left[\frac{1}{p} + \frac{1-q}{p+1} x + \cdots + \frac{(1-q)(2-q)\cdots(n-q)}{n!(p+n)} x^n + \cdots \right] \tag{T2.6b}$$

$$B(x;\, p,\, 4) = x^p \left[\frac{1}{p} - \frac{3x}{p+1} + \frac{3x^2}{p+2} - \frac{x^3}{p+3}\right] \tag{T2.7}$$

Yield calculation

$$Y = \frac{F W_\infty N_a e^{Zr + M(a-t_0)}}{K} B\left(e^{-Kr}; \frac{Z}{K}, b+1\right) \tag{T2.8}$$

$$Y|_{b=3} = F W_\infty N_a e^{M(a-t_0-r)} \left[\frac{1}{Z} - \frac{3e^{-Kr}}{Z+K} + \frac{3e^{-2Kr}}{Z+2K} - \frac{e^{-3Kr}}{Z+3K}\right] \tag{T2.9}$$

$$l_t = L_\infty \left(1 - e^{-K(t-t_0)}\right) \tag{4}$$

for length l_t with an exponential weight–length relationship

$$w_t = c l_t^b. \tag{5}$$

In (4), fish theoretically have length 0 at age t_0 and grow to asymptotic length L_∞ at a rate determined by the parameter K. The exponent b in (5) relates to allometric growth, where the body shape and density can vary with age. If fish grow isometrically (i.e., shape and density remain constant), then $b=3$. The constant c scales length to weight, where (4) and (5) imply (T2.3) with $W_\infty = cL_\infty^b$.

Suppose that regulations determine the fishing mortality rate F and the recruitment age $t_R \geq a$. The latter might be implemented by setting a size limit l_{t_R} calculated from (4). Then the fishery captures fish with age $t \geq t_R$ at the rate FN_t and biomass at the rate Fw_tN_t. Consequently, total yield Y from the cohort is given by the integral (T2.5). Assumptions (T2.1)–(T2.5) allow us to compute this integral analytically, where it simplifies matters slightly to define r in (T2.4) as the time interval between ages t_0 and t_R. The result (T2.8) depends on the incomplete beta function defined as an integral in (T2.6a), with a corresponding power series calculation (T2.6b). This handy theorem allows a fishery scientist to calculate yield, although its early application required a table of the beta function, such as that produced by Wilimovsky & Wiklund (1963).

The case $b=3$ of isometric growth gives the simpler expression (T2.9) for Y. To understand this, notice in (T2.8) that the parameter q in the beta function corresponds to $b+1$; thus, $q=4$

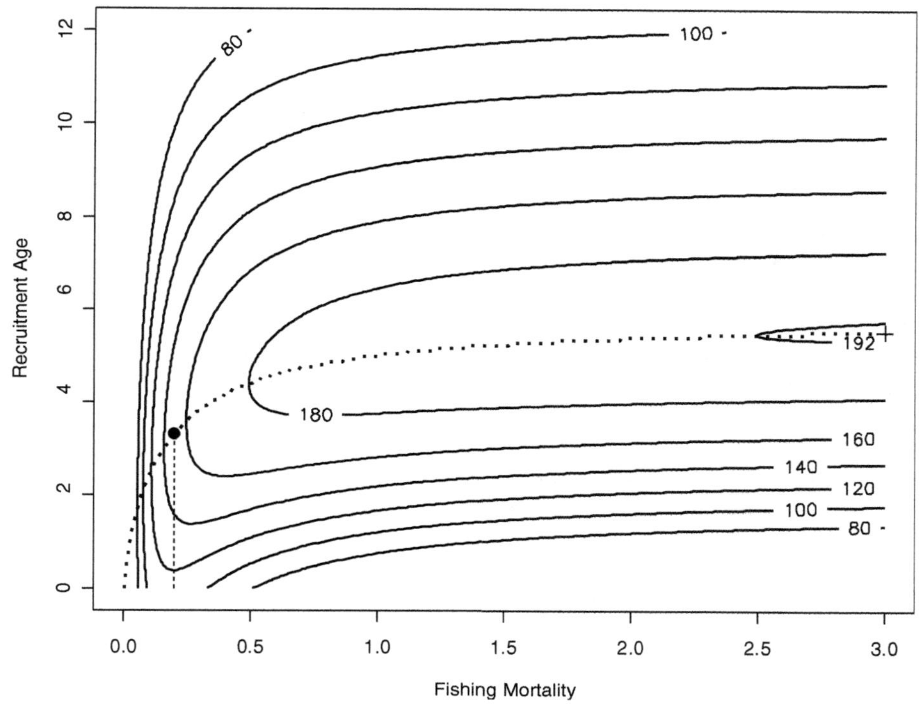

Figure 2. Yield (g) per fish at age $a = 1$ when $M = 0.2\,\text{y}^{-1}$, $W_\infty = 1209\,\text{g}$, $K = 0.2\,\text{y}^{-1}$, $t_0 = -1.066\,\text{y}$, and $b = 3$. Contours computed from (T2.8) depend on the fishing mortality rate F and the recruitment age t_R. A dotted curve shows the *eumetric* line, defined by the age t_R that gives maximum yield for a specified F. Two eumetric points correspond to $F = M = 0.2$ (•) and $F = 3$ (+). Ricker (1975, p. 254, Figure 10.2) presents a similar example, using parameters for North Sea haddock reported by Beverton & Holt (1956).

when $b = 3$. In this case, all terms involving x^n in the power series (T2.6b) vanish when $n \geq 4$, so that $B(x; p, 4)$ reduces to (T2.7). Applying this result to (T2.8) gives (T2.9).

Historically, the formula (T2.9) when $b = 3$ was discovered before the general result (T2.8). In his second Green Book, Ricker (1958, pp. 220–222; equation (10.18)) attributes (T2.9) to Beverton & Holt (1956, 1957), along with earlier literature dating back to Graham (1952). A footnote (Ricker 1958, p. 222) also mentions a result by Jones (1957) applicable even when $b \neq 3$ in the growth law (T2.3). Probably this appeared too late for inclusion in that printing of the handbook, but the 3rd edition (Ricker 1975) presents two results (pp. 251–254, equation (10.21); p. 255, equation (10.23)) that correspond, respectively, to my equations (T2.9) and (T2.8). Notation varies between the second and third handbooks, and mine is closer to that in the third. I have slightly generalized the results to allow a specified base age a for the initial population N_a. The equations cited by Ricker (1975) assume that $a = t_0$, although some of his worked examples require adjustment to a different base age. (I caution readers of the historical literature to note that the notation N_0 usually denotes the theoretical number of fish at age t_0, not age $t = 0$.)

The computation in Table 2 achieves the synthesis that Ricker (1948) foresaw in his first Green Book. Biological statistics on mortality, growth, and recruitment imply a definite value for the yield produced by a cohort; that is:

$$(F, M; W_\infty, K, t_0, b; a, N_a, t_R) \Rightarrow Y.$$

The specified age a corresponds to an initial baseline population N_a, but regulations determine the actual recruitment age $t_R \geq a$ for legal capture. Because yield Y is proportional to N_a in (T2.9), we can also regard the formula as a tool for calculating the yield per fish at age a:

$$(M, a, W_\infty, K, t_0, b; F, t_R) \Rightarrow \frac{Y}{N_a}. \tag{6}$$

where the parameters have been rearranged to distinguish fish biology (M, a, W_∞, K, t_0, b) from management policy (F, t_R). The relationship (6) allows managers to evaluate the consequences of choosing a particular fishing mortality F and age of first capture t_R.

Figure 2 illustrates a classical yield analysis, following an example from Ricker (1975). Beverton & Holt (1957) referred to contours of the yield surface as *isopleths*, and they defined the *eumetric* line by the recruitment age t_R that maximizes yield for a given fishing mortality F. At a eumetric point, the yield contour has a vertical tangent line that corresponds to the associated value F. In the example here, fish grow to weight $W_\infty = 1209$ g, but the maximum yield from a fish at the base age $a = 1$ is about 192 g. To achieve this, fish must be allowed to grow until about age $t_R = 5.5$, before exposing them to capture by the fishery at an extremely high fishing rate $F = 3.0$ y^{-1} (point marked "+"). This policy effectively crops the population at age 5.5. A more conservative plan might set $F = M = 0.2$ and harvest fish at the earlier age 3.3 (point marked "•") to achieve a yield of 152 g per fish at age 1. In fact, at this level of F, the yield contours are almost vertical, and not much yield would be lost by a precautionary increase of t_R to age 4 or 5. Such thought experiments demonstrate the power of this analysis for biological *interpretation*, the word added to the title of Ricker (1975). Essentially, (T2.8) provides a simulation model for exploring the consequences of management actions, based on the biology of mortality and growth.

Ricker (1975, pp. 254–255) mentions Fortran programs to conduct the calculations (T2.8) and (T2.9). Apparently the era of beta function tables had passed by the mid-1970s. Nevertheless, he expresses skepticism about the new technology when he suggests checking the general formula by Jones (my equation (T2.8)) with the exponent $b = 3$ to see if it gives the same result as the Beverton–Holt formula (my equation (T2.9)). In Ricker's view (*op. cit.*, p 255), "This applies even if the work has been done by computer; indeed, it is especially necessary then." He also comments (*op. cit.*, p. 257) that producing a graph like Figure 2 here "is quite tedious, because the contour lines must be interpolated among the contour values." While writing the code for Figure 2, I followed Bill's advice and checked that both model formulations generate the same results if $b = 3$. I watched with satisfaction as current technology created a grid of $200 \times 200 = 40\,000$ function values and interpolated the contours almost in the blink of an eye. But I remember how intimidating such a calculation seemed in the 1970s, with much slower computers, awkward software, and very limited graphical capabilities.

And beyond

In the year 1975 when Bill Ricker published his third handbook, Bill Gates and Paul Allen founded their now famous company Microsoft. Computing changed rapidly, with production of the Apple II computer in 1977 and the IBM counterpart in 1981. Fishery scientists could soon automate most of Bill's analyses with comparative ease on personal computers. The new technology made it possible to address questions that previously seemed intractable. For example, although a Peterson marking experiment (Table 1) could be subjected to statistical analysis, what about the yield calculation in Table 2? The mortality M and growth parameters (W_∞, K, t_0, b) must all be measured with error, just like the estimate $\hat{N}$ from a Peterson experiment. Furthermore, the so-called biological "laws" (T2.2)–(T2.3) operate stochastically at best, introducing process error. How would the analysis of Table 1 change if we introduced reasonable levels of measurement and process error?

Answers to such questions require not only better computing power, but also more general statistical theory. One approach comes from bootstrap methods dating back to the 1970s. For example, Efron (1979b) wrote a paper with the provocative title: "Computers and the theory of statistics: thinking the unthinkable." Could computers actually change theory? In the preface to their textbook published 14 years later, Eron & Tibshirani (1993) eloquently claim that:

> "Statistics is a subject of amazingly many uses and surprisingly few effective practitioners. The traditional road to statistical knowledge is blocked, for most, by a formidable wall of mathematics. Our approach here avoids that wall. The bootstrap is a computer-based method of statistical inference that can answer many real statistical questions without formulas. Our goal in this book is to arm scientists and engineers, as

well as statisticians, with computational techniques that they can use to analyze and understand complicated data sets."

The bootstrap belongs properly to the frequentist approach to statistics. According to this school of thought, parameters (like the population N in Table 1) actually exist in nature and estimates (like $\hat{N}$) have a distribution that allows us to assess variability (like the variance estimate (T1.5)). In practice, the bootstrap circumvents formulas like (T1.5) by using a computer to draw samples from the observed data and generate an empirical distribution of estimates. Historically, we benefit from the work of statisticians who obtained analytical results for specialized problems. But similar results may not be available for more complex statistical models, and the bootstrap holds promise as a tool with much greater generality.

Another general approach comes from Bayesian statistics, which uses probability to describe a subjective understanding of nature. Starting from an initial understanding (called a prior distribution), new data give a revised perspective (the posterior distribution). Parameters (like N in Table 1) become random variables whose distribution depends on prior knowledge, plus all available data. A sample drawn from the posterior reflects our current understanding of nature. In the last two decades, numerous algorithms for posterior sampling have been proposed and used. Like bootstraps, these involve computer intensive methods that can deal with very general statistical models. Clifford (1993, p. 53) described their practical impact in dramatic terms:

> "...from now on we can compare our data with the model that we actually want to use rather than a model which has some mathematically convenient form. This is surely a revolution..."

For example, consider the yield calculation in Table 2. As mentioned above, the complicated model (T2.1)–(T2.5) does not lend itself easily to measurement and process error. More precisely, given any reasonable error model, an analytical formula for the variance $V[Y]$ would be hard to obtain. The Bayesian approach, however, treats Y as a random variable with a distribution that cascades from the inputs. A posterior sample of Y, along with all random inputs, gives the analyst a sense of variability, where contour graphs like Figure 2 could portray mean values, variances, coefficients of variation, and other statistics.

In summary, the nearly three decades since Ricker's (1975) third Green Book have brought huge changes in technology and theory. As a result, the fishery biological system he described has been investigated with increasing depth. Quinn (2003) gives a readable modern perspective, and Schnute (1994, 2003) provides further background on contemporary fishery models. Ricker's foundation lies at the core of all modern work. His search for a valid interpretation of biological statistics from a fish population has set the stage for generations to come.

An inquiring mind

Cyrillids

Bill brought a compelling curiosity to any subject that interested him. For example, he introduced me to the controversies about crop circles with the investigative article by Nickell & Fischer (1992). We co-authored just one paper, and it had nothing whatever to do with fisheries. Bill's interest in the natural world extended to celestial events, such as the dramatic appearance of fireballs over eastern Canada on 9 February 1913. The astronomer Chant (1913a, 1913b) of the University of Toronto compiled anecdotal records and used "the formulas of spherical trigonometry" to determine an approximate trajectory. Later, O'Keefe (1968) found other historical reports, including sightings off northeastern Brazil. He dubbed the fireballs *cyrillids* because of their appearance on St. Cyril's Day.

Bill's contacts in the Nanaimo Historical Society led him to a story about a celestial event witnessed over Nanoose, British Columbia (about 20 km northwest of Nanaimo) during the early 1900s. Could this have been a cyrillid sighting? He stopped me in the parking lot one evening with the question: "Do you know anything about spherical trigonometry?" It was my job to extend the cyrillid trajectory westward and determine its proximity to Nanoose. We even did a trivial error analysis to take some account of uncertainty. In the end, we decided that the Nanoose report had a reasonable chance of extending the 1913 cyrillid event, rather like a range extension for a species. Fortunately, the Editor (Francis Cook) and reviewers at *The Canadian Field-Natualist* gave our speculations a sympathetic hearing, and the paper found its way into print (Ricker & Schnute 1999). To my considerable surprise while writing this current paper, a meteor streaked across the skies of northwestern North America at about 2:30 AM on 3 June 2004. Momentary brightness appeared on the record of video surveillance cameras monitoring parking lots in Portland, Oregon. Witnesses from British Columbia, Washington, and Oregon reported direct sightings, sometimes accompanied with sounds. I thought of Chant in 1913 as I listened to a modern astronomer (David Dodge of the H.R. Macmillan Space Centre in Vancouver, British Columbia) compile anecdotal reports during a local radio program.

Sunflowers

As a mathematician, I had the opportunity to participate in some of Bill's mathematical recreations. He would occasionally pass me articles of interest, such as Stewart's (1995) discussion of flower structures. What algorithm could possibly explain the beautifully tight packing of seeds on the head of a sunflower? The article appeared in the January 1995

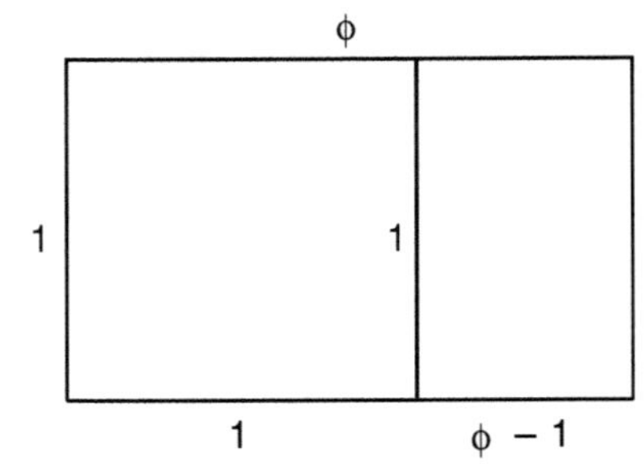

Figure 3. Rectangle illustrating the golden ratio ϕ . The outer rectangle with sides ϕ and 1 can be partitioned into a 1×1 square and an inner rectangle with sides 1 and ϕ−1. The two rectangles have the same proportion between long and short sides: $\phi/1 = 1/(\phi - 1)$.

issue of Scientific American, on sale in December 1994, so it gave us something amusing to think about during the season before Christmas. One approach to the question involves the concepts presented in Table 3. The golden ratio ϕ has a geometric definition (Figure 3) based on a rectangle with certain ideal proportions. This implies the mathematical definition (T3.1), a quadratic equation with two solutions: $\phi = (\sqrt{5}+1)/2$ in (T3.2) and $\phi - 1 = (\sqrt{5}-1)/2$. The Fibonacci numbers F_n

$$1,\ 1,\ 2,\ 3,\ 5,\ 8,\ 13,\ 21,\ 34, \ldots, \tag{7}$$

Table 3. The golden ratio ϕ defined in Figure 3, Fibonacci numbers F_n, and the packing algorithm used to produce Figure 4.

Golden ratio

$$\frac{\phi}{1} = \frac{1}{\phi - 1} \tag{T3.1}$$

$$\phi = \frac{\sqrt{5}+1}{2} \tag{T3.2}$$

Fibonacci numbers

$$F_1 = 1,\ F_2 = 1,\ F_n = F_{n-1} + F_{n-2} \tag{T3.3}$$

$$F_n = \frac{\phi^n - (1-\phi)^n}{2\phi - 1} \tag{T3.4}$$

$$\phi = \lim_{n \to \infty} \frac{F_{n+1}}{F_n} \tag{T3.5}$$

Packing algorithm and golden angle

$$(x_k, y_k) = \sqrt{\frac{k}{n}}(\cos k\theta,\ \sin k\theta), k = 1, \ldots, n \tag{T3.6}$$

$$\theta_g = \frac{360^\circ}{\phi^2} = 137.507764\ldots^\circ \tag{T3.7}$$

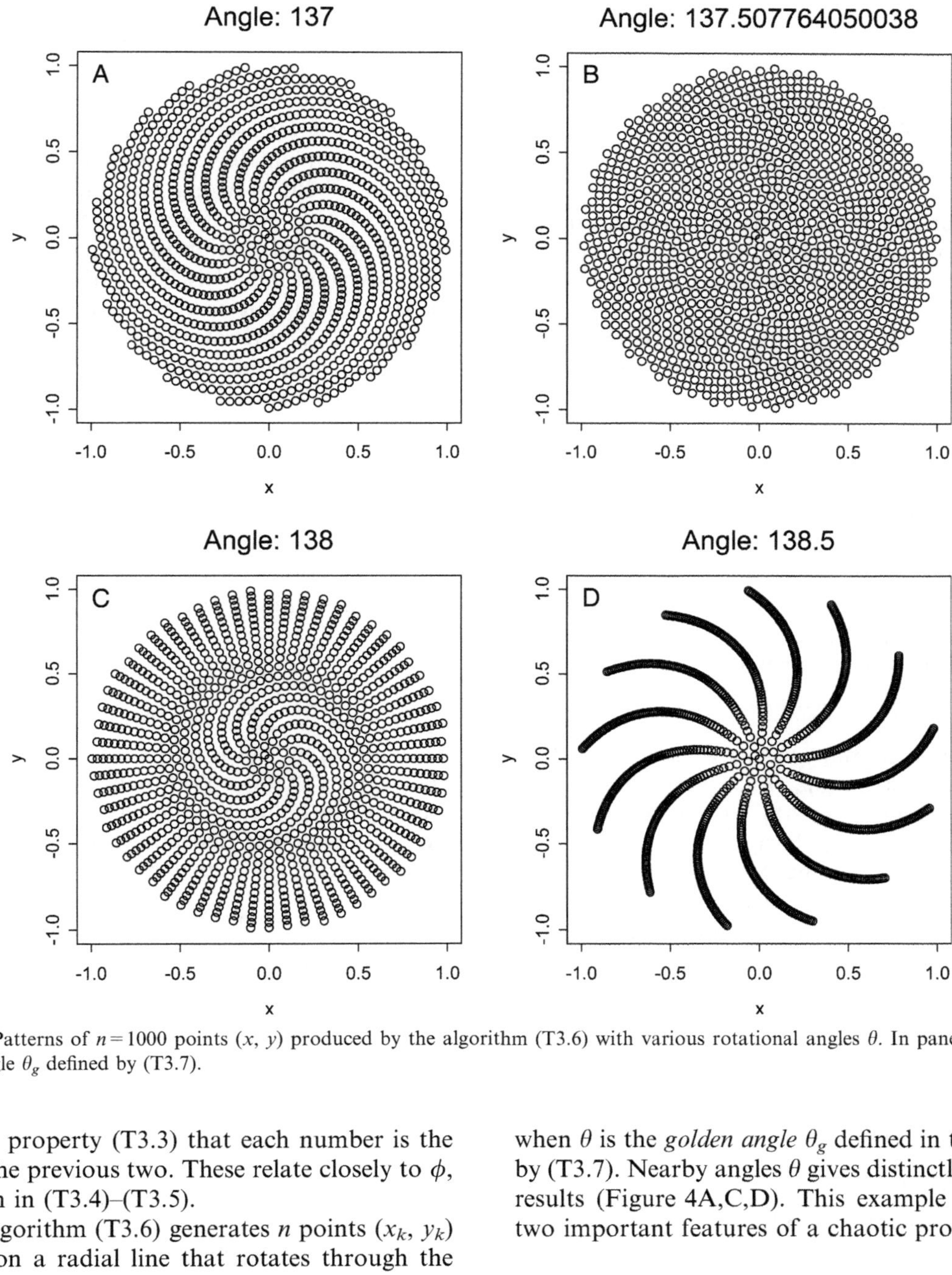

Figure 4. Patterns of $n = 1000$ points (x, y) produced by the algorithm (T3.6) with various rotational angles θ. In panel B, θ is the golden angle θ_g defined by (T3.7).

have the property (T3.3) that each number is the sum of the previous two. These relate closely to ϕ, as shown in (T3.4)–(T3.5).

The algorithm (T3.6) generates n points (x_k, y_k) located on a radial line that rotates through the angle θ at each successive point. The radial distance $\sqrt{k/n}$ increases proportional to the square root of k, giving an even spacing of points within the area of a unit circle. Figure 4 illustrates patterns produced by this algorithm for four choices of the angle θ. A structure remarkably similar to that of packed sunflower seeds (Figure 4B) occurs when θ is the *golden angle* θ_g defined in terms of ϕ by (T3.7). Nearby angles θ gives distinctly different results (Figure 4A,C,D). This example illustrates two important features of a chaotic process:

- a simple algorithm can produce a complex pattern, and
- a small change in the algorithm can produce a major change in the pattern.

I produced Figure 4 as a Christmas present for Bill in 1994.

Recruitment patterns

Ricker's famous recruitment curve first appeared in the Green Books (Ricker 1958, p. 237, equation (11.6)) in the form

$$F = Pe^{(P_r - P)/P_m}, \tag{8}$$

where F denotes the filial generation (recruitment) and P the parental generation. The model (8) has two parameters defined by Ricker (*op. cit.*) as:

- P_r the "replacement" size of the parental generation, i.e., that which, on the average, just replaces its own numbers;
- P_m the level of parental stock which produces the maximum filial generation.

In modern notation (8) has become (Ricker 1975, p. 282, equation (11.9))

$$R = \alpha S e^{-\beta S}, \tag{9}$$

where R (recruitment) replaces F, S (stock) replaces P, and the two parameters are

$$\alpha = e^{P_r/P_m}, \ \beta = 1/P_m. \tag{10}$$

Ricker himself continued using P, rather than S, to denote parent stock (*op. cit.*; Ricker 2006).

The transition from (8) to (9) illustrates the perspective that history brings to current practices. Ricker originally thought of his parameters (α, β) in terms of stock sizes that produce replacement and maximal recruitment. We can turn (10) around to compute reference points S_r and S_m (analogous to P_r and P_m):

$$S_r = \frac{\log \alpha}{\beta}, \ S_m = \frac{1}{\beta}. \tag{11}$$

A short calculation from (9) and (11) verifies the replacement equation

$$S_r = \alpha S_r e^{-\beta S_r} \tag{12}$$

and the derivative condition $dR/dS = 0$ for a maximum when $S = S_m$.

In 1995, Bill asked me why his curve (9) appeared as an example in a technical book by the Russian author Kuznetsov (1995, p. 112). Bill had a longstanding interest in Russian contributions to science, as mentioned almost casually in his autobiographical sketch (Ricker 2006): "Rumours of a major work by F.I. Baranov ... were intriguing enough that I got hold of a copy in Indiana and learned enough Russian to read it and eventually translate it." Perhaps he discovered the Kuznetsov reference because of the Russian connection, although I never asked him about that. Similarly, Kuznetsov might have noticed the Ricker curve because of Bill's strong scientific reputation in Russia. Either way, Kuznetsov's book makes a remarkable contribution to the literature on nonlinear dynamical models. The author wrote it while in Amsterdam, but he cites work at the Research Computing Centre of the Russian Academy of Sciences, which was renamed in 1992 as the Institute of Mathematical Problems in Biology.

The answer to Bill's question about Kuznetsov pertains to the replacement value S_r, which dates back to P_r in the 1958 handbook. In fact, this fundamental concept needs revision to account for some surprisingly complex behaviour of the Ricker function. Consider a sequence of stock sizes S_t determined recursively by

$$S_{t+1} = \alpha S_t e^{-\beta S_t}, \tag{13}$$

where the recruitment from one generation becomes parent stock for the next. Comparing (12) and (13), we might expect that S_t converges to the replacement value S_r, but this is not always the case. The outcome depends on critical values for the parameter α (Kuznetsov 1995, p. 114)

$$\begin{aligned} \alpha_1 &= e^2 = 7.38907\cdots, \quad \alpha_2 = 12.50925\cdots, \\ \alpha_3 &= 14.24425\cdots, \quad \alpha_4 = 14.65267\cdots, \\ \alpha_5 &= 14.74212\cdots, \ldots, . \end{aligned} \tag{14}$$

For $\alpha \le \alpha_1$, the stock has a stable replacement value S_r given by (11). However, the stock alternates between two equilibrium sizes when $\alpha_1 < \alpha \le \alpha_2$, and more generally oscillates through 2^k sizes when $\alpha_k < \alpha \le \alpha_{k+1}$.

In the remaining discussion, assume for simplicity that $\beta = 1$. (This amounts only to a change of dimensional units, because β acts as a scale parameter in the model (9).) Figure 5 illustrates the sequential process (13) as a cycle around the Ricker curve. By reflection through the replacement line where $R = S$, the recruitment from one step becomes stock for the next. When $\alpha = 7$ (Figure 5A), just below the first critical value α_1,

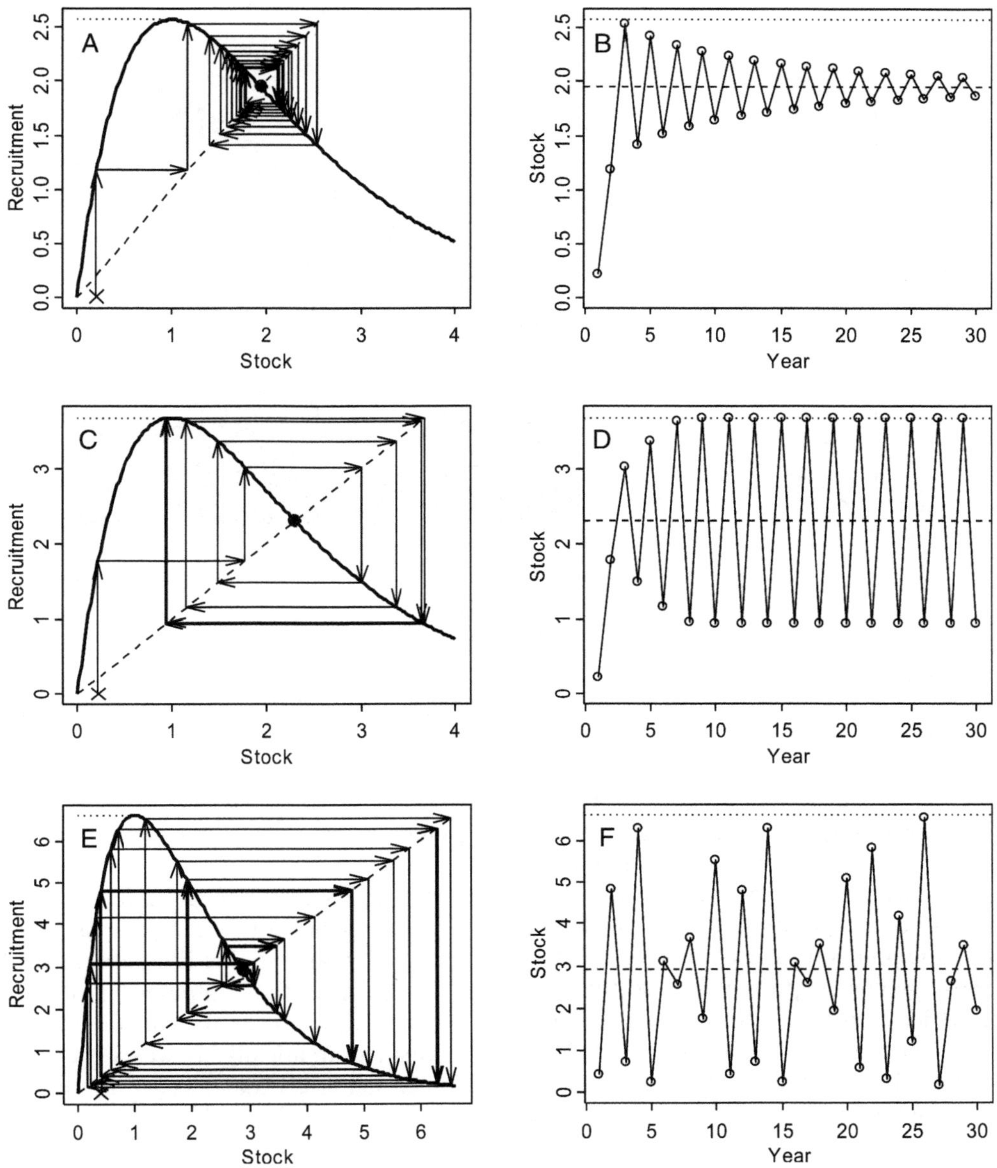

Figure 5. Recursive iteration (13) of stock sizes S_t from a Ricker curve with $\beta = 1$ and (A, B) $\alpha = 7$, (C, D) $\alpha = 10$, (E, F) $\alpha = 18$. Panels on the left show the Ricker curve (solid), the replacement line $R = S$ (broken), the replacement point • (S_r, S_r), and the trajectory's initial point ×. Arrows indicate the progression of points, where movement to the replacement line $R = S$ converts recruitment from one generation into stock for the next. Panels on the right show the resulting time series S_t, where horizontal lines indicate the levels S_r (broken) and R_{max} (dotted).

the trajectory spirals inward around the replacement point (S_r, S_r), giving a series of stock sizes that oscillate while converging toward S_r (Figure 5B). When $\alpha = 10$ (Figure 5C), between critical values α_1 and α_2, the trajectory spirals away from (S_r, S_r), producing a series of alternating stock sizes above and below S_r (Figure 5D). When $\alpha = 18$ (Figure 5E), beyond the critical values α_i in (14), the trajectory behaves erratically, and the resulting sequence moves through high and low values that do not repeat precisely (Figure 5F).

Because the replacement value S_r in (11) is not always stable, it makes sense to examine *settled*

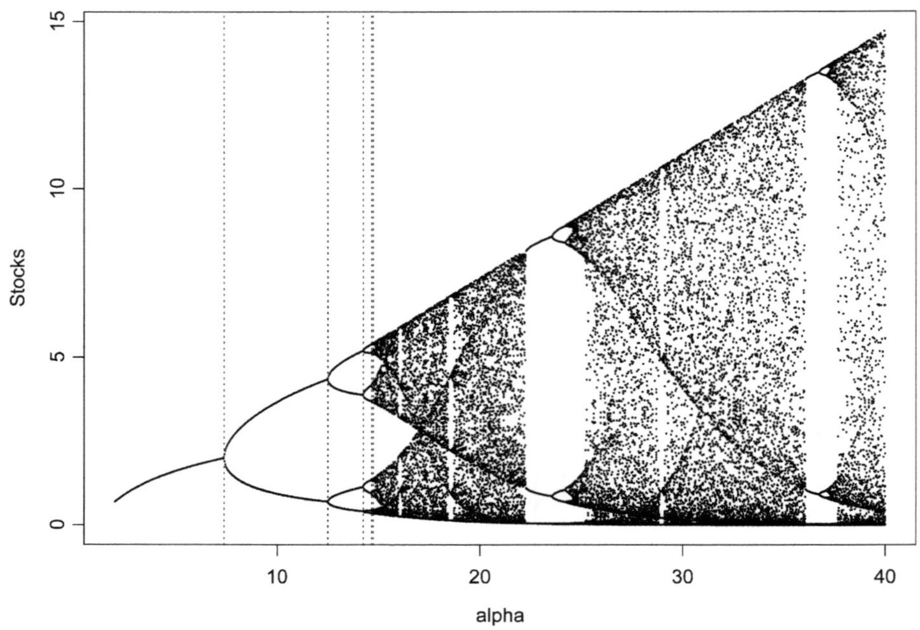

Figure 6. Bifurcation diagram of settled stock sizes in relation to α for a Ricker curve with $\beta = 1$. For each of 1000 values α ($2 \leq \alpha \leq 40$), the figure records 40 stock sizes S_t that occur after an initial trajectory of 500 iterations from (13). Vertical dotted lines indicate bifurcation values α_i listed in (14). The single-valued curve on the left ($2 \leq \alpha \leq \alpha_1$) corresponds to the replacement value $S_r = \log \alpha$ calculated from (11).

stock values that occur after a long initial trajectory. The relationship between α and settled values from (13) gives the remarkable pattern in Figure 6. Vertical lines indicate the values α_i in (14) at which the number of settled points doubles, which are called *bifurcation* points for the parameter α. They follow a nearly geometric progression, so that the number of settled points effectively becomes infinite and the Ricker curve produces chaos at values α slightly higher than α_5. Still higher values α give additional regions of regularity, bifurcation, and chaos.

The chaotic behaviour of simple models illustrated in Figures 4 and 6 came as a surprise to scientists in the 1970s, when Bill wrote his third handbook. May (2002) expresses his own mood at that time by quoting lines from a play: (Stoppard 1993, Act I, Scene 4): "It's the best possible time to be alive, when everything you know is wrong." Chaos theory undermines the orderly, predictable vision of a Newtonian world, where planets move with great regularity. The boundary between determinism and chance becomes blurred. May (*op. cit.*, p. 29) even challenges the role of computers in these developments:

> "Contrary to the suggestion you often hear that chaos theory was a computer-generated discovery, all that's needed is a paper, pencil and a lot of patience; the computer simply increases the speed with which we can do the calculations, albeit dramatically. It was with just these low-tech materials that I began my work on chaos, and in those days all I ever had to hand was the early desktop machines, which seem antediluvian by today's standards."

Ever the naturalist, Bill wanted to get a real sunflower head and compare its structure with the theoretical image in Figure 4B. Sadly, we never did this together. But his interest in Figures 4–6 illustrates the spirit of inquiry that remained part of his life long after he produced his handbooks. Like May (1976, 2002), he found motivation in ideas, not merely technology. His quest to understand biological systems remains timeless, whether it involves recruitment curves, incomplete beta functions, or sunflowers.

Acknowledgements

Gordon Miller, George Pattern, and Brian Krishka helped me assemble historical materials used in writing this paper. I also relied on Karl Ricker's (2001) list of references to his father's work. Paul Rankin kindly loaned me his personal copy of Ricker (1971). My brother David Schnute helped me recall events from the 1940s. Nick Boers assisted with graphical software in the statistical language R. Rowan Haigh and Laura Richards provided helpful suggestions for improving the text.

References

Bailey, N.J.J. 1951. On estimating the size of mobile populations from recapture data. Biometrika 38: 293–306.

Beverton, R.J.H. & S.J. Holt. 1956. A review of methods for estimating mortality rates in fish populations, with special reference to sources of bias in catch sampling. Rapport P.-V. Reunion Conseil Permenant International Exploration de le Mer 140: 67–83.

Beverton, R.J.H. & S.J. Holt. 1957. On the dynamics of exploited fish populations. U.K. Ministry of Agriculture, Fish & Fisheries Investigations (Ser. 2) 19: 533 pp.

Chant, C.A. 1913a. An extraordinary meteoric display. J. R. Astron. Soc. Can. 7: 145–215.

Chant, C.A. 1913b. Further information regarding the meteoric display of February 9, 1913. J. R. Astron. Soc. Can. 7: 438–447.

Chapman, D.G. 1951. Some properties of the hypergeometric distribution with application to zoological sample censuses. Univ. Calif. Publ. Stat. 1: 131–160.

Clifford, P. 1993. Discussion on the meeting on the Gibbs sampler and other Markov chain Monte Carlo methods. J. R. Stat. Soc. (Series B) 55: 53–102.

Efron, B. 1979b. Computers and the theory of statistics: Thinking the unthinkable. SIAM Rev. 21: 460–480.

Efron, B.E. & R.J. Tibshirani. 1993. An introduction to the bootstrap. Mongraphs on Statistics and Applied Probability, Chapman & Hall, New York, 436 pp.

Fisher, R.A. 1925. Statistical Methods for Research Workers. Oliver & Boyd, Edinburgh. 239 pp.

Fisher, R.A. 1930. The Genetical Theory of Natural Selection. Clarendon Press, Oxford. 272 pp.

Fisher, R.A. 1935. The Design of Experiments. Oliver & Boyd, Edinburgh. 252 pp.

Fisher, R.A. 1937. The Design of Experiments. 2 ed. Oliver & Boyd, Edinburgh. 260 pp.

Graham, M. 1952. Overfishing and optimum fishing. Rapport P.-V. Reunion Conseil Permanent International Exploration de le Mer 132: 72–78.

Jones, R. 1957. A much simplified version of the fish yield equation. Doc. No. P. 21, presented at the Lisbon joint meeting of International Committee on Northwest Atlantic Fisheries, International Council for the Exploration of the Sea, and Food Agriculture Organization, United Nations, 8 pp.

Kuznetsov, Y.A. 1995. Elements of applied bifurcation theory. Applied Mathematical Sciences, Vol. 112, Springer-Verlag, New York, 515 pp.

May, R.M. 1976. Simple mathematical models with very complicated dynamics. Nature 261: 459–467.

May, R.M. 2002. The best possible time to be alive. pp. 28–45. *In*: G. Farmelo, (ed.), It must be Beautiful: Great Equations of Modern Science, Granata Books, London. 284 pp.

Mood, A.M. & F.A. Graybill. 1963. Introduction to the Theory of Statistics. 2 ed. McGraw-Hill, New York. 443 pp.

Nickell, J. & J.F. Fisher. 1992. The crop-circle phenomenon: An investigative report. Skept. Inquirer 16: 136–149.

O'Keefe, J.A. 1968. New data on cyrillids. J. R. Astron. Soc. Can. 62: 97–98.

Pearson, K. 1948. Tables of the Incomplete Beta-function. The University Press, Cambridge, U.K.. 494 pp.

Petersen, C.G.J. 1896. The yearly immigration of young plaice into the Limfjord from the German Sea, etc. Rep. Danish Biol. Stat. 6: 1–48.

Quinn, T.J. II 2003. Ruminations on the development and future of population dynamics models in fisheries. Nat. Res. Model. 16: 341–392.

Ricker, K. 2001. The scientific and literary publications of W.E. Ricker, O.O.C., Ph.D., D. Sc. LLD, F.R.S.C. (1929–2001 and still ongoing). Fisheries & Oceans Canada, Pacific Biological Station Library. Bound manuscript.

Ricker, W.E. 1948. Methods of estimating vital statistics of fish populations. Indiana University Publications, Science Series, No. 15. 101 pp.

Ricker, W.E. 1958. Handbook of computations for biological statistics of fish populations. Bulletin of the Fisheries Research Board of Canada, No. 119. 300 pp.

Ricker, W.E. (ed.) 1968. Methods for assessment of fish production in freshwaters. IBP Handbook No. 3. Blackwell Scientific Publications, Oxford, 313 pp.

Ricker, W.E. (ed.) 1971. Methods for assessment of fish production in freshwaters. IBP Handbook No. 3, 2nd edition. Blackwell Scientific Publications, Oxford, 348 pp.

Ricker, W.E. 1975. Computation and interpretation of biological statistics of fish populations. Bulletin of the Fisheries Research Board of Canada, No. 191. 382 pp.

Ricker, W.E. 2006. One man's journey through the years when ecology came of age. Environ. Biol. Fishes 75: 7–37.

Robson, D.S. & H.A. Regier. 1964. Sample size in Petersen mark-recapture experiments. Trans. Amer. Fish. Soc. 93: 215–226.

Schnute, J.T. 1994. A general framework for developing sequential fisheries models. Can. J. Fish. Aquat. Sci. 51: 1676–1688.

Schnute, J.T. 2003. Designing fishery models: A personal adventure. Nat. Resour. Model. 16: 393–413.
Stewart, I. 1995. Mathematical recreations: Daisy, daisy, give me your answer, do. Sci. Am. 195: 96–99.
Stoppard, T. 1993. Arcadia. Faber & Faber, London. 97 pp.
Wilimovsky, N.J. & E.C. Wiklund. 1963. Tables of the incomplete beta function for the calculation of fish population yield. Institute of Fisheries, University of British Columbia, Vancouver, Canada. 291 pp.

Environmental Biology of Fishes (2006) 75:111–117
DOI 10.1007/s10641-005-2445-8

Bill Ricker's entomological contributions

Geoffrey G.E. Scudder
Department of Zoology, University of British Columbia, 6270 University Boulevard, Vancouver, BC, Canada V6T 1Z4 (e-mail: scudder@zoology.ubc.ca)

Received 7 June 2004 Accepted 17 June 2004

Key words: W.E. Ricker, entomology, contributions, activities, stoneflies, Plecoptera

Synopsis

Bill Ricker was an internationally recognized expert on the stoneflies (Plecoptera). These insects have aquatic larvae that live mostly in cool and clear running water. They are of ecological significance in the breakdown of leaf-litter and the cycling of detritus in aquatic lotic habitats. Since they can account for a significant portion of the diet of some fishes in cold northern waters, they are linked with Ricker's other work in fisheries. Within 1 year of taking up the study of stoneflies, Ricker was describing species new to science. He continued such activity through most of his life, describing or co-describing some 108 species and 46 genera. Ricker made other outstanding contributions, publishing in 1952 what is regarded as the most important publication dealing with stonefly systematics in the 20th century. Many names he coined in his studies clearly demonstrated his linguistic flair and humour. He made significant contributions to an understanding of the biogeography of stoneflies, and received many honours and citations for his research.

William E. (Bill) Ricker is best known in entomology for his outstanding and pioneer studies on the stoneflies (Plecoptera). However, this group of insects, with aquatic larvae, was not the first group of choice in his insect studies. In his childhood, Bill Ricker showed an early interest in natural history, with his first passion being the study of plants and birds (Ricker 2006). Although largely self-taught, he became adept at their identification to species. This flair was to characterize all his later taxonomic studies.

It was in 1927, during his second year as an undergraduate at the University of Toronto, that he took an invertebrate course from the dragonfly expert, E.M. Walker. It is not surprising to find that he soon started studying these insects. Walker and other faculty members encouraged him in his taxonomic investigations, and during the summers, employed him to analyze the food content of trout stomachs, and carry out ecological appraisals of trout-ponds and streams. He quickly became engrossed in freshwater biology, and his ongoing entomological studies soon focussed on dragonflies (Odonata) and caddisflies (Trichoptera), which he sought to identify to species.

It was not until 1934, on the suggestion of E.M. Walker, and his colleague Fred Ide, who was his companion during 3 years of stream and lake study in Ontario in 1928–1930 (Ricker 1992), that he started to concentrate on the stoneflies (Plecoptera). From then onwards, this Order became the main focus of his taxonomic investigations. This was quite logical, as aquatic stonefly larvae are of ecological significance in freshwater ecology, because they breakdown leaf-litter and are important in the cycling of detritus in freshwater lotic habitats. Since stonefly larvae can account for a significant portion of the diet of some fishes in cold northern waters, they readily linked with Bill's other work and interest in fisheries. Also, because these aquatic insects live mostly in cool and clear running water, they were ideal subjects to study for someone who soon was to begin his abiding love of the west coast of Canada (Gordon Hartman and Neil Bourne in Stewart (2002)).

Bill started off concentrating on the Plecoptera in British Columbia in the Cultus Lake basin and surrounding areas, as part of his Ph.D. studies, supervised by E.M. Walker. Specimen data show that he also made collecting trips to Douglas Lake and Nicola in 1934–1935, Subsequently, he collected stoneflies wherever he traveled over the years. He soon became an internationally recognized expert on the group, having studied material in European collections in the winter of 1935–1936 (Ricker 1938).

Within 1 year of deciding to concentrate on the Plecoptera, Ricker published two papers describing species new to science in this group (Ricker 1935a, b). Altogether, between 1935 and his last paper on stoneflies, with K.W. Stewart on the Plecoptera of the Yukon (Stewart & Ricker 1997), he authored or co-authored 34 papers (or chapters) on the group. They are listed in chronological order in Stewart (2002). There was an additional paper with his former Ph.D. supervisor E.M. Walker, on the dragonflies from the vicinity of Cultus Lake (Walker & Ricker 1938). Prior to this, Walker (1937) had named a new dragonfly species after Ricker that the latter had collected at Cultus Lake (Table 4).

Ricker's own paper on the stoneflies from the vicinity of Cultus Lake (Ricker 1939), listed 52 species, mostly of his own collecting. His later paper on the stoneflies of southwestern British Columbia (Ricker 1943) considered 80 species. Ricker & Scudder (1975) increased the British Columbia list of stoneflies to 120 species. Since then, only five additional species have been added (Cannings 1989).

Exactly half of the papers Ricker authored or co-authored on stoneflies, contain the description of new taxa. In total this involved some 108 new species and 46 new genera (or subgenera) (Ricker 1992).[1]

[1] Ricker (1992) actually listed 47 genera, including *Setvena* Ricker 1952, but this genus is now attributed to Illies (1966) because the latter first designated the type species.

Examination of the 617 stoneflies listed from North America by Stark (1997) shows that there are 105 species described by Ricker alone or with others (Table 1): 86 are still valid.

Ricker was very much involved in the resurgence of the study of North American stoneflies in the 1930s, and developed a keen interest in their higher classification, evolution and biogeography. He (Ricker 1943) started splitting the larger genera of Plecoptera into subgenera, and in this paper erected seven of these new taxa. Many other such taxa were erected in his ground-breaking monograph entitled 'Systematic studies in Plecoptera' (Ricker 1952). In essence, he completely rearranged some parts of the classification of the Order, mainly on the basis of the evolutionary development of the genitalia. Many plecopterists regard this 1952 paper as the most important publication dealing with stonefly systematics to be published in the 20th century (Bill Stark, in litt). Ricker showed remarkable insight and ability to recognize natural groupings and relationships among those groups in a broad array of taxa. Others have noted that 'what he built from the wreckage was a thing of beauty and simplicity, that made evolutionary sense' (Boyle 1984).

Most of the subgenera erected by Ricker have subsequently been raised to generic level, either by Illies (1966) or Baumann (1975). As a result, of the 102 genera of stoneflies in North America, over one-third or 35 (Table 2) are Ricker's taxa: only two of his additional names are in synonymy. Ricker's (1952) study, and work with H.H. (Herb) Ross (Ricker & Ross 1975), also resulted in the recognition of new taxa from elsewhere in the world, and he had earlier (Ricker 1935b) described a new fossil species from Prussian Baltic amber (Table 3). Collectively, this is a phenomenal contribution to systematic entomology.

Bill's interest in the evolution and biogeography of the Plecoptera, is perhaps best illustrated by reference to his 1964 paper (Ricker 1964), and his monograph with H.H. Ross on the winter stonefly genus *Allocapnia* (Ross & Ricker 1971). This latter work as the authors noted in the Introduction, started from some informal discussion with colleagues in 1956, about which insect groups have distribution patterns that might contribute reliable information concerning the effects of the Pleistocene glaciers on the biota of eastern North

Table 1. List of species of North American Plecoptera with Bill Ricker as author or co-author. Current family placement in parentheses, with any new generic placement after this citation. Those names in italics are synonyms of the taxon listed after the family name.

agassizi Ricker 1943, Isocapnia (Capniidae)
appalachia Ricker & Ross 1975, Strophopteryx (Taeniopterygidae)
arkansae Ricker & Ross 1975, Strophopteryx (Taeniopterygidae)
arnoldi Ricker & Ross 1969, Zealeuctra (Leuctridae)
atlanticum Ricker & Ross 1975, Taenionema (Taeniopterygidae)
aurora Ricker 1952, Allocapnia (Capniidae)
baddecka Ricker 1965, Leuctra (Leuctridae)
bergi Ricker 1965, Capnia (Capniidae) Mesocapnia
beringi Ricker 1965, Capnia (Capniidae) Mesocapnia variabilis (Klapálek 1920)
besametsa Ricker 1952, Nemoura (Prostoia) (Nemouridae) Prostoia
burksi Ricker & Ross 1968, Taeniopteryx (Taeniopterygidae)
calcarea Ricker 1935, Hastaperla (Chloroperlidae) Haploperla brevis (Banks 1895)
cheama Ricker 1965, Capnia (Capniidae)
chila Ricker 1952, Nemoura (Zapada) (Nemouridae) Zapada
chilnualna Ricker 1952, Hastaperla (Chloroperlidae) Haploperla
concolor Ricker 1935, Alloperla (Chloroperlidae)
cotta Ricker 1952, Isoperla (Perlodidae)
cuestae Ricker 1935, Acroneuria (Perlidae) Acroneuria carolinensis (Banks 1905)
cunninghami Ross & Ricker 1971, Allocapnia (Capniidae)
delosa Ricker 1952, Nemoura (Amphinemura) (Nemouridae) Amphinemura
dusha Ricker 1965, Paraleuctra (Leuctridae) Paraleuctra occidentalis (Banks 1901)
fattigi Ricker 1949, Paragentina (Perlidae) Paragentina kansensis (Banks 1905)
foersteri Ricker 1943, Nemoura (Nemouridae) Ostrocerca
fosketti Ricker 1965, Brachyptera (Oemopteryx) (Taeniopterygidae) Oemopteryx
fraseri Ricker 1943, Diploperla (Perlodidae) Cultus aestivalis (Needham & Claassen 1925)
fraseri Ricker 1959, Isocapnia (Capniidae)
fraxina Ricker & Ross 1969, Zealeuctra (Leuctridae)
frisoni Ross & Ricker 1964, Allocapnia (Capniidae)
fumosa Ricker 1935, Acroneuria (Perlodidae) Paragentina immarginata (Say 1823)
gregsoni Ricker 1965, Capnia (Bolshecapnia) (Capniidae) Bolshecapnia
hansoni Ricker 1952, Isogenus (Isogenoides) (Perlodidae) Isogenoides
hantzschi Ricker 1938, Capnia (Capniidae) Capnia nearctica Banks 1918
haysi Ricker 1952, Nemoura (Zapada) (Nemouridae) Zapada
hitei Ricker & Ross 1969, Zealeuctra (Leuctridae)
hubbsi Ricker 1952, Neoperla (Perlidae)
hyalita Ricker 1959, Isocapnia (Capniidae)
idei Ricker 1935, Chloroperla (Chloroperlidae) Alloperla
inaya Ricker & Ross 1975, Strophopteryx (Taeniopterygidae) Strophopteryx appalachia Ricker & Ross 1975
indianae Ricker 1952, Allocapnia (Capniidae)
krumholzi Ricker 1952, Isogenus (Isogenoides) (Perlodidae) Isogenoides
labradora Ricker 1951, Capnia (Capniidae) Utacapnia
laurie Ricker 1952, Peltoperla (Peltoperlidae) Tallaperla
leonarda Ricker 1952, Alloperla (Chloroperlidae)
linda Ricker 1952, Nemoura (Amphinemura) (Nemouridae) Amphinemura
lonicera Ricker & Ross 1968, Taeniopteryx (Taeniopterygidae)
loshada Ricker 1952, Allocapnia (Capniidae)
macdunnoughi Ricker 1948, Nemoura (Nemouridae) Podmosta
mariana Ricker 1943, Peltoperla (Peltoperlidae) Yoraperla
medveda Ricker 1952, Alloperla (Chloroperlidae)
metequi Ricker & Ross 1968, Taeniopteryx (Taeniopterygidae)
milnei Ricker 1935, Chloroperla (Chloroperlidae) Alloperla chloris Frison 1934
missourii Ricker 1959, Isocapnia (Capniidae)
mockfordi Ricker 1952, Nemoura (Amphinemura) (Nemouridae) Amphinemura
mogila Ricker 1959, Isocapnia (Capniidae)
moha Ricker 1952, Leuctra (Leuctridae)
mohri Ross & Ricker 1964, Allocapnia (Capniidae)

Table 1. Continued.

narfi Ricker & Ross 1969, Zealeuctra (Leuctridae)
normani Ricker 1952, Nemoura (Nemouridae)
ohioensis Ross & Ricker 1964, Allocapnia (Capniidae)
okanagan Ricker 1935, Acroneura (Perlidae) Hesperoperla pacifica (Banks 1900)
onkos Ricker 1935, Alloperla (Chloroperlidae) Sweltsa
ostra Ricker & Ross 1975, Strophopteryx (Taeniopterygidae) Strophopteryx cucullata Frison 1934
ovibovis Ricker 1965, Chloroperla (Chloroperlidae) Alaskaperla
pechumani Ross & Ricker 1964, Allocapnia (Capniidae)
peltoides Ross & Ricker 1964, Allocapnia (Capniidae)
perplexa Ross & Ricker 1964, Allocapnia (Capniidae)
pintada Ricker 1952, Alloperla (Triznaka) (Chloroperlidae) Triznaka
polemistis Ross & Ricker 1971, Allocapnia (Capniidae)
rogozera Ricker 1965, Capnia (Bolshecapnia) (Capniidae) Bolshecapnia
rossi Ricker 1952, Nemoura (Podmosta) (Nemouridae) Podmosta decepta (Frison 1942)
salvelini Ricker 1935, Acroneuria (Perlidae) Paragentina media (Walker 1852)
sandersoni Ricker 1952, Allocapnia (Capniidae)
sasquatchi Ricker 1965, Capnia (Bolshecapnia) (Capniidae) Bolshecapnia
scotti Ricker 1952, Pteronarcys (Pteronarcyidae)
smithi Ross & Ricker 1971, Allocapnia (Capniidae)
sopladora Ricker 1952, Utaperla (Chloroperlidae)
spenceri Ricker 1965, Capnia (Bolshecapnia) (Capniidae) Bolshecapnia
spenceri Ricker 1943, Isocapnia (Capniidae)
sugluka Ricker 1965, Capnia (Capniidae) Mesocapnia
tamalpa Ricker 1952, Alloperla (Sweltsa) (Chloroperlidae) Sweltsa
tennessa Ross & Ricker 1964, Allocapnia (Capniidae)
thalia Ricker 1952, Alloperla (Chloroperlidae) Alloperla severa (Hagen 1861)
thujae Ricker 1943, Isocapnia (Capniidae) Isocapnia spenceri Ricker 1943
tina Ricker 1952, Nemoura (Malenka) (Nemouridae) Malenka
torontonensis Ricker 1935, Allocapnia (Capniidae) Allocapnia pygmaea (Burmeister 1839)
tostonus Ricker 1952, Isogenus (Cultus) (Perlodidae) Cultus
townesi Ricker 1952, Alloperla (Sweltsa) (Chloroperlidae) Sweltsa
tumana Ricker 1952, Nemoura (Nemouridae) Lednia
ugola Ricker & Ross 1968, Taeniopteryx (Taeniopterygidae)
urticae Ricker 1952, Alloperla (Sweltsa) (Chloroperlidae) Sweltsa
usa Ricker 1952, Alloperla (Chloroperlidae)
varshava Ricker 1952, Nemoura (Amphinemura) (Nemouridae) Amphinemura
vedderensis Ricker 1943, Eucapnopsis (Capniidae) Isocapnia
vershina Gaufin & Ricker 1975, Paraleuctra (Leuctridae)
voinae Ricker 1948, Alloperla (Chloroperlidae)
vostoki Ricker 1948, Alloperla (Chloroperlidae)
wachita Ricker & Ross 1969, Zealeuctra (Leuctridae)
walkeri Ricker 1943, Arcynopteryx (Perlodidae) Frisonia picticeps (Hanson 1942)
warreni Ricker & Ross 1969, Zealeuctra (Leuctridae)
watertoni Ricker 1952, Arcynopteryx (Megarcys) (Perlodidae) Megarcys
weberi Ricker 1952, Nemoura (Podmosta) (Nemouridae) Podmosta
wenatchee Ricker 1965, Nemoura (Malenka) (Nemouridae) Malenka
wilsoni Ricker 1965, Paraperla (Chloroperlidae)
zelona Ricker 1965, Brachyptera (Oemopteryx) (Taeniopterygidae) Oemopteryx fosketti (Ricker 1965)
zola Ricker 1952, Allocapnia (Capniidae)

America. They remark that early in the discussion it became clear that highly vagile insects, such as winged grasshoppers, and leafhoppers disperse so rapidly that their present distribution is largely an expression of ecological rather than historic factors. Certain wingless insects, such as the Collembola or springtails, have remarkably wide ranges, are easily transported by air currents and therefore are not valid as expressions of historic dispersals within a limited area. They continue, 'At this point the idea of the cold climates associated with the Pleistocene glaciers, became associated in

Table 2. List of genera of North American Plecoptera with Bill Ricker as author or co-author. Current family placement in parentheses. Names in italics are synonyms of the taxon listed after the family name.

Attaneuria Ricker 1955 (Perlidae)
Bolotoperla Ricker & Ross 1975 (Taeniopterygidae)
Bolshecapnia Ricker 1965 (Capniidae)
Calineuria Ricker 1955 (Perlidae)
Chernokvilus Ricker 1952 (Perlodidae)
Cultus Ricker 1952 (Perlodidae)
Despaxia Ricker 1943 (Leuctridae)
Dolkrila Ricker 1952 (Perlodidae) Diura Billberg 1820
Frisonia Ricker 1943 (Perlodidae)
Hastaperla Ricker 1935 (Chloroperlidae) Haploperla Navas 1934
Helopicus Ricker 19452 (Perlodidae)
Kogotus Ricker 1952 (Perlodidae)
Lednia Ricker 1952 (Nemouridae)
Malenka Ricker 1952 (Nemouridae)
Malirekus Ricker 1952 (Perlodidae)
Moselia Ricker 1943 (Leuctridae)
Neaviperla Ricker 1943 (Chloroperlidae)
Osobenus Ricker 1952 (Perlodidae)
Ostrocerca Ricker 1952 (Nemouridae)
Podmosta Ricker 1952 (Nemouridae)
Prostoia Ricker 1952 (Nemouridae)
Rasvena Ricker 1952 (Chloroperlidae)
Remenus Ricker 1952 (Perlodidae)
Shipsa Ricker 1952 (Nemouridae)
Skwala Ricker 1943 (Perlodidae)
Soliperla Ricker 1952 (Peltoperlidae)
Soyedina Ricker 1952 (Nemouridae)
Suwallia Ricker 1943 (Chloroperlidae)
Sweltsa Ricker 1943 (Chloroperlidae)
Triznaka Ricker 1952 (Chloroperlidae)
Utaperla Ricker 1952 (Chloroperlidae)
Viehoperla Ricker 1952 (Peltoperlidae)
Visoka Ricker 1952 (Nemouridae)
Yoraperla Ricker 1952 (Peltoperlidae)
Yugus Ricker 1952 (Perlodidae)
Zapada Ricker 1952 (Nemouridae)
Zealeuctra Ricker 1952 (Leuctridae)

our minds with the cold-tolerant stoneflies'. A rapid check of some of their distributions as known at that time turned up a most interesting item. The winter stonefly *Allocapnia pygmaea* (Burmeister) was primarily a subboreal species in northeastern and eastern North America, but it had an isolate population in the Ozark Mountain area of Missouri. The Missouri population appeared to be a segment of a cold-adapted species left stranded to the south when the remainder of the species moved northward in post-glacial times. 'Our interest, was caught by this circumstance so

Table 3. Stonefly genera (subgenera) and species described by Bill Ricker and collaborators from outside North America.

Besdolus Ricker 1952 (Perlodidae)
Isocapnia kudia Ricker 1959 (Capniidae)
Kohnoperla Ricker & Ross 1975 (Taeniopterididae)
Megaleuctra neavei Ricker 1935 (fossil) (Leuctridae)
Mesyatsia Ricker & Ross 1975 (Taeniopterididae)
Okamotoperla Ricker & Ross 1975 (Taeniopterididae)
Ostrovus Ricker 1952 (Perlodidae)
Sopkalia Ricker 1952 (Perlodidae)
Stavsolus Ricker 1952 (Perlodidae)
Tadamus Ricker 1952 (Perlodidae)
Tadamus kohnonis Ricker 1952 (Perlodidae)
Zhiltzovia Ricker & Ross 1975 (Taeniopterygidae) syn. of *Brachyptera* Newport 1849

we decided to study the winter stoneflies with special reference to Pleistocene events'.

They noted that the winter stoneflies belong to a physiologically peculiar group of organisms, namely ones in which winter heralds not a cessation of growth and activity as in most insects, but instead an acceleration of growth and activity. They went on to study these insects further, aided, as they said, by the 'Winter Stonefly Club', a group like ourselves, that enjoyed getting out for a little brisk collecting when the desk chairs in the office begin to harden in January, February and March. They showed that when the geographical distribution was integrated with the phylogeny, dispersal paths for present species and hypothetical ancestral species could be adduced.

They demonstrated that the genus *Allocapnia* apparently evolved primarily in association with the Appalachian Mountain System, its neighbouring ridges, and areas northeast to them. They postulated that six ancestral lineages spread to the Ozark–Ouachita Mountain region, but not synchronously. The speciation pattern of *Allocapnia* was associated with the alternation of cold glacial and warm interglacial periods of the Pleistocene and comparable climatic oscillations occurring in the late Pliocene. The genus as we know it today was said to be 3 or 4 million years old.

Bill had picked up some of these same ideas and approaches in an earlier paper on the 'Distribution of Canadian Stoneflies' that he presented in September 1963 at the Third International Symposium on Plecoptera, held in Plön, Germany. In the printed paper (Ricker 1964) he stated that to interpret recent stonefly distributions in North America, two maps need to be kept in mind. One, a map of present physiography and vegetation, the other, a map of ice cover during the Pleistocene.

He presented a number of maps depicting the distribution of stoneflies, analyzed the patterns shown within the 214 Canadian species then known, and noted that the Canadian distribution patterns cannot be usefully considered apart from those of the same species in the United States. He proceeded to interpret these again in terms of Pleistocene and post-Pleistocene events, and was particularly intrigued by the disjuncts, what he called the rare and unusual species. It is of interest to note that many of Bill's ideas have been substantiated by more recent research in palynology, and studies on the distribution of fossil insects, especially the arctic and cold-adapted subarctic beetles.

Others have commented on Ricker's facility with foreign languages (Boyle 1984). In Plecoptera circles, Bill's linguistic flair is evident in some of the names he used in describing his new stonefly taxa. While most taxonomists base their names, as is customary, on Latin or Greek, Bill preferred to use Spanish, native American or Russian words for his scientific names, noting that in coining the immense number of names now in existence, the classical languages have been rather thoroughly ransacked (Ricker 1992).

Ricker (1992) has explained the origin of the stonefly names that he proposed alone or with collaborators, and his sense of humour is obvious. Thus, the species name *usa*, as in *Alloperla usa*, a species he named in 1952, comes from the Russian *us* = moustache. He chose it because of the patch of setae this stonefly species has on its epiproct. Another species he named *Zapada chila*. The generic name *Zapada* that he coined in 1952, comes from the Russian *západ*, meaning west, because the genus occurs in western North America. The species named *chila*, comes from the Spanish for red pepper. Ricker chose this particular name also in 1952 because he thought that this particular insect was a 'red hot discovery', as it was the first *Zapada* species found in the east, in Tennessee. It is not so hot as he thought because another species, *Zapada kathdin* Baumann & Mingo has since been discovered in Maine.

Ricker (1992) also explains that he named *Alloperla* (*Sweltsa*) *urticae* after *urtica* = nettle, since he ran into some while collecting this species. He also points out that the use of the species name *zola*, as in *Allocapnia zola* Ricker 1952, comes from the Russian *zolá* = ashes (of a fire). He stated (Ricker 1992) that the types were from Ash Cove, Ohio, 'which is a very poor pun that I never expected to divulge to anyone'.

Over the years, Ricker had many taxa named after him (Table 4), and he continued his earnest collection and identification of stoneflies until microscope work was no longer possible owing to his failing eyesight, for close-up work, in the 1990s. Bill's attention then returned to his early passion–plants (Ricker 2006), and he donated his personal stonefly collection, most of it in alcohol, to the Canadian National Collection of Insects, housed in Agriculture and Agri-Food Canada in Ottawa.

However, Bill remained actively interested in plecopterology, and in contact with other plecopterists right up until his death (Stewart 2002). He attended the XI International Symposium on Plecoptera in Treehaven in 1992, the XII International Symposium in Lausanne in 1995, and the North American Plecoptera Symposium in Montreal in 1997 (Stewart 2002). In 1992 he received the Lifetime Achievement Award from the International Society of Plecopterists, and in 1995, the Rolling Stonefly News Award from the North American Plecoptera Society. These he added to the many other honours and awards, received throughout this life.

As noted by Boyle (1984), William E. Ricker's stature in stonefly and fisheries research was so outstanding that some scientists automatically assumed that there were two experts working in two different fields, with the same name and same

Table 4. Some insects named after Bill Ricker.

Order ODONATA
Macromia rickeri Walker 1937 (Macromiidae)
Order PLECOPTERA
Allocapnia rickeri Frison 1942 (Capniidae)
Hydroperla rickeri (Stark 1984) (Perlodidae)
Leuctra rickeri James 1976 (Leuctridae)
Nemoura rickeri Jewett 1971 (Nemouridae)
Paraleuctra rickeri Nebeker & Gaufin 1966 (Leuctridae)
Rickera Jewett 1954 (Perlodidae)
Order TRICHOPTERA
Neophylax rickeri (Milne 1935) (Uenoidae)

middle initial. Many are still surprised to discover that there was only one William E. (Bill) Ricker, and that he was indeed, an expert in both fields.

Acknowledgements

I have greatly benefited by being able to consult the summaries, accounts, tables and lists of Bill's research compiled by his son Karl, and I thank him for making these available. I am indebted to Mary Arai, R.W. Baumann, E.C. Becker, R. Beamish, Doug Currie, H.V. Danks, Rex Kenner, Bill Stark and Ken Stewart for help with the literature and my queries. Launi Lucas not only processed the manuscript for this paper, but also sought out some of the references for me when I was away from Vancouver. Finally, I thank David Noakes for sending me copies of Karl Ricker's articles printed herein, and asking me to contribute to this volume commemorating Bill's lifetime achievements.

References

Baumann, R.W. 1975. Revision of the stonefly Family Nemouridae (Plecoptera): A Study of the World Fauna at the Generic Level. Smithsonian Contributions to Zoology 211. 74 pp.

Boyle, R.H. 1984. Fisheries biologist William Ricker is a real Hall of Famer in his field. Sports Illustrated, November 12, 1984.

Cannings, S.G. 1989. New records of slender winter stoneflies (Plecoptera: Capniidae) in British Columbia. J. Entomol. Soc. Brit. Columbia 86: 77–80.

Illies, J. 1966. Katalog der rezenten Plecoptera Das Tierreich 82. Walter de Gruyter & Co, Berlin. 662 pp.

Ricker K.E. 2006. Flora was his interest and prime course of study: a botanical career for W.E. Ricker disappears. Environ. Biol. Fish. 75: 39 – 72.

Ricker, W.E. 1935a. Description of three new Canadian perlids. Can. Entomol. 67: 197–201.

Ricker, W.E. 1935b. New Canadian perlids (Part 2). Can. Entomol. 67: 256–264.

Ricker, W.E. 1938. Notes on specimens of American Plecoptera in European collections. Trans. Royal Can. Inst. 22(1): 129–156.

Ricker, W.E. 1939. A preliminary list of stoneflies (Plecoptera) from the vicinity of Cultus Lake, British Columbia. Proc. Entomol. Soc. Brit. Columbia 35: 19–23.

Ricker, W.E. 1943. Stoneflies of southwestern British Columbia. Indiana University Publications, Science Series No. 12. 145 pp.

Ricker, W.E. 1952. Systematic studies in Plecoptera. Indiana University Publications, Science Series No. 18. 200 pp.

Ricker, W.E. 1964. Distribution of Canadian stoneflies. Gewässer und Abwässer 34/35: 50–71.

Ricker, W.E. 1992. Origin of stonefly names proposed by Ricker and collaborators. Perla 18: 1–12.

Ricker W.E. 2006. One man's journey through the time when ecology came of age. Environ. Biol. Fishes 75: 7 – 37.

Ricker, W.E. & H.H. Ross. 1975. Synopsis of the Brachypterinae (Insects, Plecoptera, Taeniopterygidae). Can. J. Zoo. 53: 132–153.

Ricker, W.E. & G.G.E. Scudder. 1975. An annotated checklist of the Plecoptera (Insecta) of British Columbia. Syesis 8: 333–348.

Ross H.H. & W.E. Ricker. 1971. The Classification, Evolution and Dispersal of the Winter Stonefly Genus *Allocapnia*. Illinois Biological Monographs 45. 166 pp.

Stark, B.P. 1997. Plecoptera. pp. 641–664. *In*: R.W. Poole & P. Gentili, (eds.), Nomina Insecta Nearctica. A Check List of the Insects of North America. Volume 4: Non-Holometabolous Orders, Entomological Information Services, Rockville, MD.

Stewart, K. 2002. Celebration of Bill Ricker's life and professional contributions to plecopterology. Perla 20: 40–49.

Stewart, K.W. & W.E. Ricker. 1997. Stoneflies (Plecoptera) of the Yukon. pp. 201–222. *In*: H.V. J.A. Danks Downes, (eds.), Insects of the Yukon, Biological Survey of Canada (Terrestrial Arthropods), Ottawa.

Walker, E.M. 1937. A new Macromia from British Columbia (Odon. Corduliidae). Can. Entomol. 69: 6–13.

Walker, E.M. & W.E. Ricker. 1938. Notes on Odonata from the vicinity of Cultus Lake, B.C. Can. Entomol. 70: 144–151.

Environmental Biology of Fishes (2006) 75:119–129
DOI 10.1007/s10641-005-2446-7

Bill Ricker: A man of gifted intellect, insatiable curiosity and generous spirit

Richard J. Beamish[a] & Donald J. Noakes[b]
[a]*Pacific Biological Station, Fisheries and Oceans Canada, 3190 Hammond Bay Road, B. C. V9T 6N7, Nanaimo, Canada (e-mail: beamishr@dfo-mpo.gc.ca)*
[b]*School of Advanced Technologies and Mathematics, Thompson Rivers University, 900 McGill Road, B.C. V2C 5N3, Kamloops, Canada (e-mail: dnoakes@tru.ca)*

Received 21 July 2004 Accepted 14 August 2004

Key words: music, mathematics, fishery, W. E. Ricker, family

Dr. Bill Ricker spent the time from his teens to his nineties understanding nature and quietly became one of the twentieth century's most influential biologists. His science was an honest communication of his observations, analyses and interpretations that allowed others to make intelligent decisions. He was always an advocate of science rather than an advocate of a particular point of view. Biological science was his passion providing him with an unending variety of adventures.

Writing about Bill Ricker inevitably is an interpretation that is made through one's own experience. Bill's scientific contributions were vast and mostly apparent through his publications. His influence, however, is immeasurable, other than to note that his name is known to all students and practitioners of fisheries science around the world. It is important to know Bill Ricker because he was a model scientist as well as a model Canadian. He made maximum use of his exceptional talents in an attempt to ensure that plants, animals, and their ecosystems were better understood and protected. Very few biologists can expect to have the talent and abilities of Bill Ricker. But all biologists should be motivated by his approach and passion for science.

It was about 3 p.m., 31 December 1994 and we were sitting in Don's office at the Pacific Biological Station enjoying a glass of New Year's Eve cheer. The Station was refreshingly quiet. There was a tapping at the partially opened office door and Dr. Ricker poked in his head. Instead of a traditional Happy New Year greeting he said, 'last chance to discover something new for this year.'

Bill Ricker had a life of discovery. To Bill, every day was a good day to discover something new. His publications include primary papers, translations, dictionaries, poetry, fiction, and topics too difficult to describe briefly (Figure 1). In fisheries, he is best known for his 1954 paper on stock and recruitment and for his Handbook of Computation for Biological Statistics of Fish Populations, which was first published in 1958 and eventually known as the 'green book.' The three editions of the green book summarized the contributions of a number of well-known fisheries scientists at a time that fisheries science and fisheries management were finding their roots around the world. Governments needed advice on how many fish were available to harvest safely and industry needed to be able to assess the economic opportunities. The handbook became a standard method for learning how to assess the impacts of fishing. The handbook was translated into Russian and Chinese and is still widely used around the world.

Dr. Ricker's famous 1954 paper simply titled 'Stock and Recruitment' resulted from his research while he taught at Indiana University from 1939 to

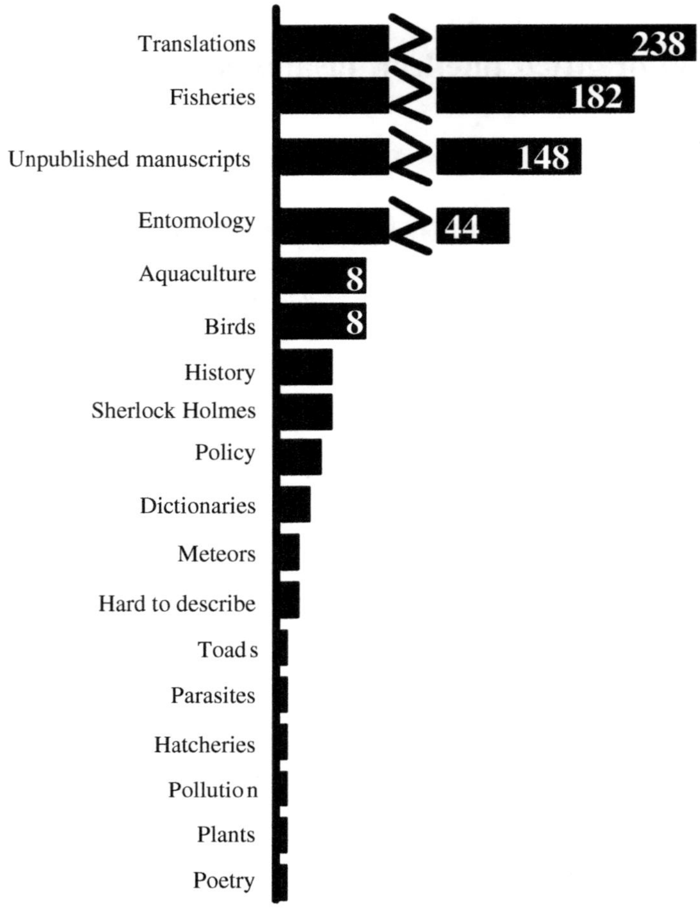

Figure 1. General topics of Bill Ricker's manuscripts and the number of publications in each category.

1950. At Indiana, he found it interesting to study the impact of sport fishing on bluegill sunfish, *Lepomis machrochirus*. He heard about the research of a famous Russian scientist named Fedor Ivanovich Baranov, which was published in 1916. In typical Ricker fashion, he learned enough Russian to translate Baranov's paper. Bill clearly credits and acknowledges the influence of the catch equation proposed by Baranov on his own work (Ricker 2005). The 1954 paper introduced the now famous 'Ricker curve,' which allows the number of juveniles (recruitment) to be estimated from the number of spawning adults (stock). Pacific salmon management had its roots in this paper. Most of what was done and is still done is based on the assumptions and results of this one paper. Although we now recognize some difficulties with the use of Ricker curves, it is most remarkable that this discovery still has such a major influence on the management of fisheries around the world some 50 years later.

This period also marked the beginning of a lifelong association between Bill and his Russian colleagues that we believe benefited both countries and fisheries science immensely. He knew that there were 10 times as many fisheries scientists in Russia working on the same issues so he learned Russian and read their publications. Bill recognized the need for open communications among scientists and his initial work with his Russian colleagues established a good foundation of mutual trust and respect. The result has been a refreshing and often vigorous exchange of scientific views with our Russian colleagues that we believe would not have been possible (at least to the same degree) without Ricker's pioneering efforts. While technological advances have made advances in certain areas of fisheries science easier or even feasible, Ricker taught us the importance of thoughtful and constructive exchanges of scientific points of view. Peer exchanges and review were always cornerstones of Ricker's success as a scientist.

Bill Ricker was also a world authority on stoneflies. There is a story that some scientists believed that there were two W.E. Ricker's, the fishery biologist and the entomologist. His interest in stoneflies included their zoogeography as well as their systematics. Bill is recognized as describing 81 of the 617 species of stoneflies currently listed in North America. In addition, one third of the 101 genera were originally identified by Bill Ricker (Scudder 2005). Fellow entomologists considered that his system of classification was 'a thing of beauty and simplicity that made evolutionary sense.'

He wrote many other papers on fish, fisheries, and fisheries management that were influential, but one paper that Dr. Ricker enjoyed producing has largely gone unnoticed. We would like to highlight this paper, because it is an example of the treasures that can be found by re-reading Dr. Ricker's publications. A wise student would benefit from reading anything Ricker wrote about a topic before stepping too deeply in to a problem. In the mid-1960s Dr. Ricker was invited to be part of a team that was put together by the United States National Academy of Science to look at the world's future natural resources. Bill's chapter was 'Food from the sea.' Other chapters included 'The human ecosystem,' 'Interactions between man and his resources,' 'Food from the land,' 'Mineral resources from the sea,' and 'Energy resources.' Bill Ricker's job was to forecast the food that could be produced by the world's oceans. At the time there was a wide belief that quantities of food available from the sea were unlimited (Huxley undated). Bill calculated that the world fish catch might be sustainable at approximately 2.5 times the 1968 catch, or about 150 to 160 million t and that the maximum world catch would be attained by the year 2000. In 2000, the world capture fisheries reported catches of approximately 95 million t with bycatch or discards accounting for an additional 30% for a total of about 125 million t (FAO 2002). Most scientists consider that this is the maximum catch or even above the maximum (Pauly et al. 1998). The accuracy of the estimate is impressive, but the method used to make the estimate is just as remarkable. One must be careful about claiming that something is first, but we believe that the chapter Bill wrote in 1968 was the first use of what is now labelled as Ecopath (Polovina 1984). Anyone using Ecopath or Ecosim (Pauly et al. 2000) should read Ricker's 1969 paper.

Rod Langley produced a video on the contribution and accomplishments of Bill Ricker that is entitled 'A passion for science.' The objective of the video was to highlight aspects of Bill Ricker's life that would be of interest to students considering a career in fisheries science. The difficulty was keeping the video to 26 min. Left out of the video was his attempt to improve sockeye salmon, *Onchorhynchus nerka*, production by removing northern squawfish, *Ptychocheilus oregonensis* (Richardson), a key predator in Cultus Lake, in British Columbia, Canada. About 80% of the adult predators were removed resulting in an initial increase in sockeye production. However, sockeye production quickly declined as the carrying capacity of the lake limited juvenile sockeye growth. At the same time, the recruitment of squawfish soared as there was little control through cannibalism. The declining production and the increased predation eventually reduced sockeye salmon abundance. Not everything Bill Ricker did turned out as planned, but everything was a learning experience. The one thing that Bill Ricker always did was to ensure that he communicated his experiences to others so that they would learn from his trials and tribulations whether successful or not.

A few years ago, Bill Ricker was asked at an official function what he would advise about the future management of Pacific salmon. He quickly responded that he had learned to 'expect the unexpected.' Pacific salmon are one of the groups of fishes in the world that receive considerable attention of fisheries biologists. They are important in many of the coastal rivers and streams flowing into the north Pacific and they dominate the surface waters of the subarctic Pacific. Pacific salmon are an icon of environmental health and general quality of life of people along the rim of

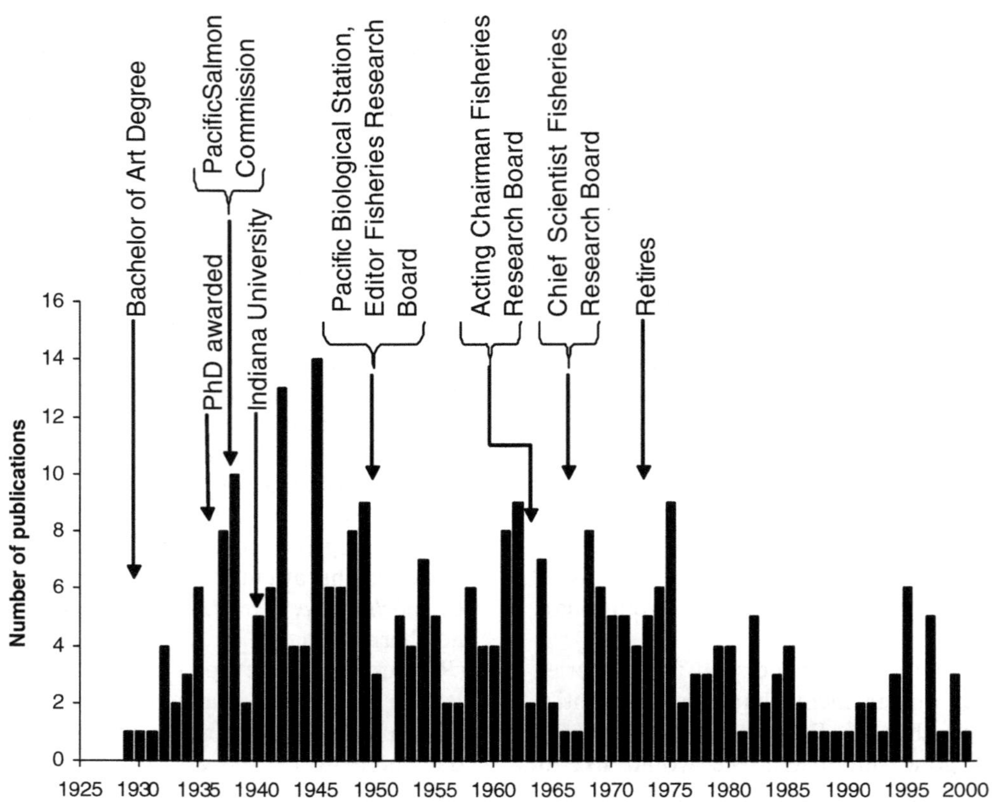

Figure 2. Chronology of Bill Ricker's major activities and the number of his annual publications up to 2000.

the Pacific from California to Busan. If the return of adult salmon to a spawning area is below expectations, someone knows and someone makes a fuss. Pacific salmon stocks are actually quite healthy despite what you might hear or read in the media. There are problems, but the generally good health of one of the world's most important groups of fishes is because Bill Ricker (and others) puzzled over the mysteries of the mechanisms that regulated their abundances. His scientific contributions have been important, but his advice to expect the unexpected should remain a fundamental rule in the management of Pacific salmon.

Bill Ricker provided a structure to manage Pacific salmon at a time that the modern fishery was developing. His standard approach for his work on Pacific salmon and science in general, was to make a series of simple calculations and as relationships became clear, more complicated computations were considered. He was always careful to clearly articulate his assumptions and his reasons for choosing a certain course of action. Biologists would be well advised today to employ the same standards. While he was careful to adopt appropriate technologies to assist him in his modeling and other work, he would always check his calculations using his favorite slide rule. It was his way of instilling a degree of common sense and logic in his solutions.

In 1973, Bill Ricker retired at the age of 65. He continued his passion for science through to early 2001 when he no longer could come to his office at the Pacific Biological Station. Bill died on 8 September 2001, shortly after his 93rd birthday. Our last conversation with Bill just days before his death was about the British Columbia fishery. He was genuinely interested in the state of stocks and the fisheries (particularly Pacific salmon) and was keen to offer his own thoughts on the current state of affairs.

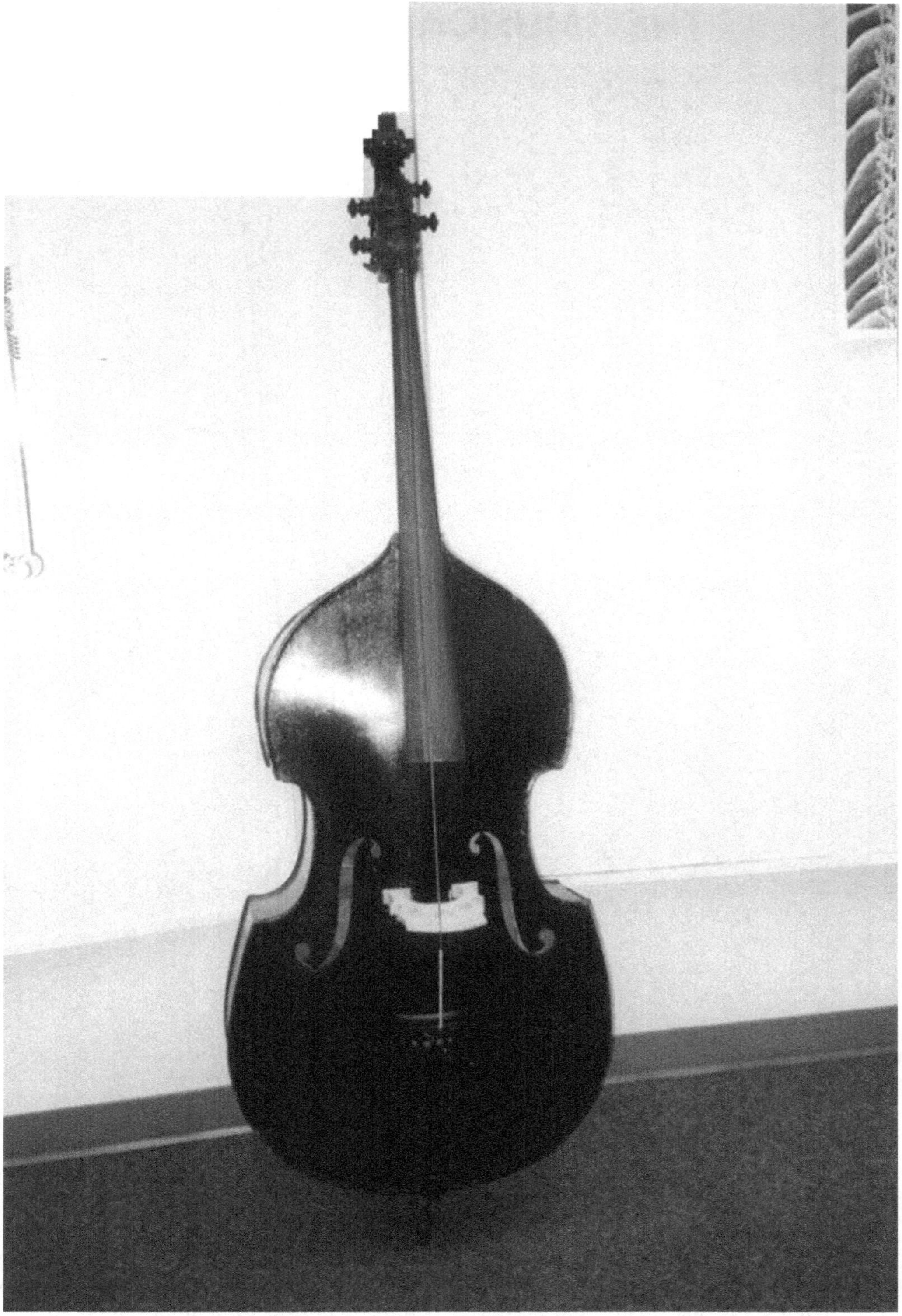

Figure 3. Bill Ricker's double bass.

THE MUSICAL RICKERS

MUSICAL FAMILY are the Rickers of Nanaimo, all of whom play with Nanaimo Symphony. With Dr. W. E. Rickers are sons, Eric (left front), John (right front) and Karl (top left). The Nanaimo orchestra plays first concert of season tonight.

Figure 4. A photograph in the Nanaimo paper early in the 1950s, showing Bill Ricker and his musical family.

It is impossible to summarize his life in this paper and even a book of normal size would be insufficient. Highlights of his travels and accomplishments are summarized in Figure 2. He finished his PhD at the University of Toronto in 1935, married Marion Torrance Cardwell, and took a postdoctoral honeymoon to Europe. He worked briefly for the, then new Pacific Salmon

Table 1. Reasons for influence of Bill Ricker.

– Gifted intellect
– Attracted to nature
– Insatiable curiosity
– Well trained in high school and university
– Skilled in quantitative analysis
– Led the development of commercial fishing with quantitative approaches
– Approaches were simple and powerful
– Elegance of insight
– Remarkable memory of literature, people, anything
– Master of the Russian fisheries literature
– Researched issues that were important to fishermen and managers
– Extraordinarily generous
– Competitive
– Dependable
– Shy but friendly; private but comfortable with himself
– Enjoyed young people
– Non-judgmental in his scientific approaches and let the data guide him to his conclusions
– Never thought of himself as being better than others
– Honest in his science
– Published

Commission in 1937–1938 in British Columbia. He left Canada in 1939 to teach at Indiana University for 12 years. J. R. Dymond managed to persuade Bill to return to Canada to head up the publications unit of the Fisheries Research Board. Bill turned a rather parochial journal into one of the world's most influential publications on aquatic sciences. He was a meticulous, but supportive editor. He described his work as an editor as sufficient to allow him to carry out research that was useful, interesting, and compatible with an office existence. As editor, Bill was also able to keep abreast of the latest research results. It is important to note that Bill did some of his best science during this period. This was not only a period of development of the fisheries off Canada's Pacific coast, but also a period of rapid expansion of world fisheries. Large fleets were being built with the expectation of finding food, employment, and profit. Governments and industry were desperate for scientific advice and Bill Ricker was now positioned to provide that assistance.

It is also important to know that Bill Ricker was a respected parent and a loving and devoted husband. Bill's sons remarked that he placed their interests first. Bill loved to do things with his sons and their friends, particularly when it involved music. In the early 1950s, two of his four sons played in the Nanaimo Symphony Orchestra. Bill had played the violin in the University of Toronto Orchestra, and learned to play the balalaika but, wanting to join his sons, he asked the conductor what instrument was needed. He learned to play the double bass (Figure 3) in 6 months and he and a third son joined the Nanaimo Symphony Orchestra (Figure 4). Bill had two songbooks that recorded his favorite songs. He apparently knew one song in more than one half dozen languages. We listened to him sing in Russian at a surprise 84th birthday party for him in a dining room of the Avachie Hotel in Petroplavlosk, Kamchatka. We also heard a recording of him singing some songs from his books only a few weeks before his death in a voice that sounded quite professional. Bill kept records of birds he had seen, plants he had found, and mountains he had climbed (Ricker 2005b). He seemed shy, but he enjoyed talking to people sharing both his knowledge and experience and people enjoyed talking to him. After his beloved wife Marion died in 1991, he would attend house parties, fitting seamlessly into an assemblage of personalities ranging from writers and artists to rugby players.

Dr. Ricker's contributions to biological sciences are recorded in his publications. His influence may be immeasurable but the reasons for his influence are worth considering. Like other intangibles, the reasons may vary among observers. The list

Table 2. Publications that Bill Ricker considered to be significant and several more that we selected.

Ricker, W.E. 1934. An ecological classification of certain Ontario streams. University of Toronto Studies, Biological Series No. 37, pp. 1–114. (Publications of the Ontario Fisheries Research Laboratory, No. 49, 1–114). [Fisheries Research Board of Canada, Studies Supplement No. 507].

Ricker, W.E. 1934. A critical discussion of various measures of oxygen saturation in lakes. Ecology 15: 348–363.

Ricker, W.E. 1937. The concept of confidence or fiducial limits applied to the Poisson frequency distribution. Journal of the American Statistical Association 32: 349–356.

Ricker, W.E. 1937. Physical and chemical characteristics of Cultus Lake, British Columbia. Journal of the Biological Board of Canada 3: 363–402.

Ricker, W.E. 1937. The food and the food supply of sockeye salmon (*Oncorhynchus nerka* Walbaum) in Cultus Lake, British Columbia. Journal of the Biological Board of Canada 3: 450–468.

Ricker, W.E. 1938. On adequate quantitative sampling of the pelagic net plankton of a lake. Journal of the Biological Board of Canada 4: 19–32.

Ricker, W.E. 1938. Seasonal and annual variations in the quantity of pelagic net plankton, Cultus Lake, British Columbia. Journal of the Biological Board of Canada 4: 33–47.

Ricker, W.E. 1938. A comparison of the seasonal growth rates of young sockeye salmon and young squawfish in Cultus Lake. Biological Board of Canada, Progress Reports of the Pacific Coast Stations No. 36: 3–5.

Ricker, W.E. 1938. "Residual" and kokanee salmon in Cultus Lake. Journal of the Fisheries Research Board of Canada 4: 192–218.

Ricker, W.E. 1941. The consumption of young sockeye salmon by predacious fish. Journal of the Fisheries Research Board of Canada 5: 293–313.

Forester, R.E. & W.E. Ricker. 1942. The effect of reduction of predaceous fish on survival of young sockeye salmon at Cultus Lake. Journal of the Fisheries Research Board of Canada 5: 315–336.

Ricker, W.E. 1942. The rate of growth of bluegill sunfish in lakes of northern Indiana. Investigations of Indiana Lakes and Streams 2: 161–214.

Ricker, W.E. 1942. Creel census, population estimates and rate of exploitation of game fish in Shoe Lake, Indiana. Investigations of Indiana Lakes and Streams 2: 215–243.

Ricker, W.E. 1945. A method for estimating minimum size limits for obtaining maximum yield. Copeia 1945: 84–94.

Ricker, W.E. 1945. Abundance, exploitation and mortality of the fishes in two lakes. Investigations of Indiana Lakes and Streams 2: 345–448.

Ricker, W.E. 1946. Production and utilization of fish populations. Ecological Monographs 16: 373–391.

Anonymous. 1947. Editors: relax, please! American Scientist 35: 306–318.

Ricker, W.E. 1948. Methods of estimating vital statistics of fish populations. Indiana University Studies, Science Series No. 15: 101 pp.

Ricker, W.E. 1950. Cycle dominance among the Fraser sockeye. Ecology 31: 6–26.

Forester, R.E. & W.E. Ricker. 1953. The coho salmon of Cultus Lake and Sweltzer Creek. Journal of the Fisheries Research Board of Canada 10: 293–319.

Ricker, W.E. 1954. Stock and Recruitment. Journal of the Fisheries Research Board of Canada 11: 559–623.

Ricker, W.E. 1955. Fish and fishing in Spear Lake, Indiana. Investigations of Indiana Lakes and Streams 4: 117–161.

Ricker, W.E. 1955. *Review*: "The natural regulation of animal numbers," by D. Lack. Journal of Wildlife Management 19: 487–488.

Ricker, W.E. 1955. *Review*: "The distribution and abundance of annuals," by H.G. Andrewartha & L.C. Birch. Journal of Wildlife Management 19: 488–489.

Ricker, W.E. 1958. Maximum sustained yields from fluctuating environments and mixed stocks. Journal of the Fisheries Research Board of Canada 15: 991–1006.

Larkin, P.A. & W.E. Ricker. 1964. Canada's Pacific marine fisheries: past performance and future prospects. pp. 194–268. *In*: P.A. Larkin & W.E. Ricker (ed.) Inventory of the Natural Resources of British Columbia, British Columbia National Resource Conference, Victoria, B.C.

Ricker, W.E. 1969. Effects of size-selective mortality and sampling bias on estimates of growth, mortality, production, and yield. Journal of the Fisheries Research Board of Canada 26: 479–541.

Ricker, W.E. 1969. Food from the sea. pp. 87–108. *In*: P. Cloud (chair.) Resources and man, a study and recommendations, Report of the Committee on Resources and Man, U.S. National Academy of Sciences, W.H. Freeman and Co., San Francisco.

Ricker, W.E. 1972. Hereditary and environmental factors affecting certain salmonid populations. pp. 19–160. *In*: A. N. Other (editor) The Stock Concept in Pacific Salmon, University of British Columbia, H.R. McMillan Lectures in Fisheries, Vancouver, B.C.

Ricker, W.E. 1973. Linear regressions in fishery research. Journal of the Fisheries Research Board of Canada 30: 409–434.

Table 2. Continued.

Ricker, W.E. 1973. Two mechanisms that make it impossible to maintain peak-period yields from stocks of Pacific salmon and other fishes. Journal of the Fisheries Research Board of Canada 30: 1275–1286.
Ricker, W.E. 1975. Computation and interpretation of biological statistics of fish populations. Bulletin of the Fisheries Research Board of Canada No. 191: 382 pp.
Ricker, W.E. & H.D. Smith. 1975. A revised interpretation of the history of the Skeena River sockeye salmon. Journal of the Fisheries Research Board of Canada 32: 1369–1381.
Ricker, W.E. 1976. Review of the rate of growth and mortality of Pacific salmon in salt water, and noncatch mortality caused by fishing. Journal of the Fisheries Research Board of Canada 33: 1483–1524.
Ricker, W.E. 1981. Changes in the average size and average age of Pacific salmon. Canadian Journal of Fisheries and Aquatic Sciences 38: 1636–1656.
Ricker, W.E. 1984. Computation and uses of central trends lines. Canadian Journal of Zoology 62: 1897–1905.
Ricker, W.E. 1987. Effects of the fishery and of obstacles to migration on the abundance of Fraser River sockeye salmon (*Oncorhynchus nerka*). Canadian Technical Report of Fisheries and Aquatic Sciences No. 1522: 75 pp.
Ricker, W.E. 1992. Back-calculation of fish lengths based on proportionality between scale and length increments. Canadian Journal of Fisheries and Aquatic Sciences 49: 1018–1026.

(Table 1) includes more than the usual 10 items and may appear too long to the reader. However, the list may be useful for those starting out in biological sciences and we would prefer to err on the long side rather than the short side. Each item is self-explanatory, but we wanted to comment on a few. In one of Richard Fyneman's books, he comments that honesty in science is telling others what they need to know to make an intelligent decision. The opposite could range from advocacy to even dishonesty, where people are told what they need to know to agree with one's views. Bill Ricker was impeccably honest, in his science and in all aspects of his life. His sons commented that his early generalist education was the foundation for his interpretative ability in his career. Ranking as the number one science student entering university in Ontario in 1926 also was noteworthy (Ricker 2005c). It was also noteworthy that Bill Ricker never boasted about his intellectual abilities. It was after his death that his family found out that Bill was a gold medalist of his class, every year at high school. It helped that Bill Ricker's career corresponded to the time that fisheries science was evolving. But the most important reason for his influence was that he published extensively. A compilation of his publications includes 296 published papers and books, 238 published translations, and 148 scientific literary manuscripts (Ricker 2005). We believe that anyone considering a career in biological science should study Bill Ricker. If your educational institution does not provide this background then consider reading the papers in Table 2. Almost all are papers that Bill identified as highlights, but a few are our favorites. Biologists of all ages will enjoy his 1955 review of the books written by the fiercely competitive Lack and Andrewartha & Birch (Table 2). The 1972 paper on hereditary and environmental factors affecting salmonid populations remains as one of the best papers on the stock concept for Pacific salmon. The Hell's Gate slide in 1913–1914 was considered to be the cause of the decline of Pacific salmon. For another perspective, read Bill Ricker's 1987 paper. Bill Ricker knew the scientific literature. He was clear in his assumptions and he wrote them down. He was meticulous and used common sense and sound logic.

Bill Ricker received approximately 34 medals and awards (Table 3). These included three honorary doctoral degrees from the University of Manitoba, Dalhousie University, and the University of Guelph. The recognition that particularly pleased him was the naming of the Canadian Fisheries Research Vessel, the W.E. Ricker in 1986, the same year he was appointed as an officer of the Order of Canada. Bill Ricker's life is an inspiration to young people contemplating a career in biological sciences. Few could expect to achieve as much as he did, but everyone could learn from his approach to science. We believe that Bill Ricker and his contributions in science are the most practical illustration of science itself. Bill Ricker was an unassuming, gentle, humble man. He is one of those giants in science on whose shoulders future biologists will stand.

Table 3. Awards, medals and honours received by Bill Ricker.

North Bay Collegiate Institute – Goldpin (Form 2 to Form 3)	
North Bay Collegiate Institute – Goldpin (Form 3 to Form 4)	
North Bay Collegiate Institute – Goldpin (Junior Matriculation)	
North Bay Collegiate Institute – Goldpin (Honours, Matriculation)	
1926	Edward Blake Scholarship for Science
1930	British Association Advancement of Science (Commemoration of Meeting in Toronto, 1924) – Bronze Medal for Advancement of Science
1930	Victoria University – First in Science Medal (Gold)
1931	University of Toronto – Masters of Arts (M. A.)
1936	University of Toronto – Ph.D.
Pacific Biological Station – Mic-a-Mac Medal	
1955	The Wildlife Society – Outstanding Fish Ecology Management Award (for 1953/54)
1956	Royal Society of Canada – Elected as a Fellow Member
1956	International Association for Theoretical and Applied Limnology – Edgardo Baldi Lectureship Award
1960	The Wildlife Society – Outstanding Fish Ecology and Management Award (for 1959)
1966	Canadian Government – Professional Institute Public Service Medal (Gold) – for Meritorious Achievement
1967	Confederation of Canada (1867–1967) – (Silver)
1969	American Fisheries Society – Award of Excellence (Gold)
Victoria University (Univ. of Toronto) – 60th Anniversary Medal	
1970	Royal Society of Canada (Regalis Societas Canadensis MDCCCLXXXII) – Flavelle Medal (Gold) for Meritous Achievement
1970	University of Manitoba – Honourary Doctor of Science (D.Sc.)
1971	Government of Canada – Certificate of Service to Government of Canada
1973	Honourary Member, Canadian Society of Zoologists
1974	Dalhousie University – Honourary Doctor of Law (LL. D.)
Classic Citation Award – for publication "Stock and Recruitment"	
1983	Canadian Society of Zoologist – F. E. J. Fry Medal (Gold)
1985	Classic Citation Award – for publication "Linear Regressions in Fishery Research"
1985	Association Professional Biologists of British Columbia – Honourary Life Membership
1986	Governor General of Canada – Officer of the Order of Canada
Bootmakers of Toronto (Canadian Holmes) – Derrick Murdoch Award	
Entomological Society of British Columbia – Honourary Life Membership	
1990	Ecological Society of America – Eminent Ecologists Award
1990	Classic Citation Award – for textbook "Computation and Interpretation of Biological Statistics of Fish Populations"
1991	Victoria University Chancellor's Council – Certificate of Appreciation
1992	Canadian Government (Governor General) – Commemorative Medal and Certificate for the 125th Confederation of Canada
1992	Confederation 125th Year Medal
1992	International Society Plectopterist – Lifetime Achievement Award in recognition of outstanding life-long work with and contribution to Plecopterology and Fisheries Biology (Plaque, issued at the XI International Plecoptera Symposium at Tomahawk, Wisconsin)
1993	National Fishing Hall of Fame (World Recognition) "Outstanding Achievement in the Realm of Fresh Water Sport Angling"
1994	American Fisheries Research Institute Award
1995	North American Plectoptera Society – Rolling Stonefly News Award, for excellence in Plecoptera Research
Vancouver Public Aquarium – Murray A. Newman Award	
1996	University of Toronto Chancellor's Circle – 65th Anniversary of Graduation Medal (1930–1995), Bronze Medal
1998	University of Guelph – Honourary Doctor of Science (D.Sc.)
1998	University of Guelph Sigma Xi – Distinguished Canadian Science Award
American Fisheries Society (North Pacific International Chapter) – Worthy Coelacanth Award	
American Fisheries Society (Marine Fisheries Section) – Oscar Elton Sette Outstanding Marine Fishery Biologist Award	
American Association for Advancement Science – 50 Year Life MemberCertificate	
2001	Government of Canada – 5 NR Science Award to Leaders in Sustainable Development (provides scholarship funds to young promising scientists in Dr. W. E. Ricker's name)

References

Food and Agriculture Organization of the United Nations (FAO). 2002. FAO Yearbook, Fishery statistics, Capture production 2000. Volume 90/1, FAO Fisheries Series 60, FAO Statistics Series 166.

Huxley, T.G. undated. Report on marine fisheries to British Government (cited in Kurlansky's book on cod).

Pauly, D., V. Christensen, J. Dalsgaard, R. Froese & F. Torres Jr. 1998. Fishing down marine food webs. Science 279: 860–863.

Pauly, D., V. Christensen & C. Walters. 2000. Ecopath, Ecosim, and Ecospace as tools for evaluating ecosystem impact of fisheries. ICES J. Mar. Sci. 57: 697–706.

Polovina, J.J. 1984. An overview of the ECOPATH model. Fishbyte 2: 5–7.

Ricker, K. 2005. Bibliography of W. E. Ricker. Environ. Biol. Fish. 74: 000–000.

Ricker, K. 2005b. Bill Ricker's records as an ornithologist. Environ. Biol. Fish. 75: 73–93.

Ricker, K. 2005c. Flora was his interest and prime course of study; a botanical career for W.E. Ricker disappears. Environ. Biol. Fish. 75: 39–72.

Scudder, G.G.E. 2005. Bill Ricker's entomological contributions. Environ. Biol. Fish. 75: 111–117.

Environmental Biology of Fishes (2006) 75:131–151
DOI 10.1007/s10641-005-2447-6

Bibliography for W.E. Ricker, B.A., M.A., Ph.D., O.C., F.R.S.C., LLD, D.Sc. March 1998

Karl E. Ricker
3253 Archibald Way, V0N 1B3, Whistler, British Columbia, Canada (e-mail: accutypeservices@shaw.ca–Attention Ricker)

Received 2 October 2004 Accepted 15 May 2005

Key words: publications, translations, manuscripts

Bill Ricker compiled his own bibliography, published in this issue. This bibliography, compiled from Bill's files and other records, is much closer to a complete listing of all of Bill's works. It is cross-referenced to Bill's version by inclusion of his numbering system, and indications of which publications he judged most significant. The Inventory #Number (KR) assigned is by Karl Ricker for this bibliography. Bill's numbers are given as WER, and correspond to the numbers in Bill's own bibliography. The Code: * indicates key publications of significance (W.E. Ricker's viewpoint). Other notations include: Date (date of publication or translation), FRB = Fisheries Research Board of Canada, S.S. = Studies Supplement, Fisheries Research Bd. of Canada. Unless indicated otherwise, author is W.E. Ricker.

1. Abridged list of Publications

KR	WER	Date	
1	1	1929	W.J.K. Harkness & W.E. Ricker. A preliminary study of some trout waters of Ontario. Transactions of the American Fisheries Society, 59: 256–265.
2	2	1930	Feeding habits of speckled trout in Ontario waters. Transactions of the American Fisheries Society, 60: 64–72.
3		1931	An ecological study of the Mad River and notes on other Ontario streams with special reference to the speckled trout. M.A. thesis, University of Toronto. 88 pp.
4	3	1932	Physiological changes and the origin of species. Canadian Field-Naturalist, 46: 30–31.
5	4	1932	Studies of speckled trout (*Salvelinus fontinalis*) in Ontario. University of Toronto Studies, Biological Series No. 36, (Publications of the Ontario Fisheries Research Laboratory, No. †44): 69–110.
6	5	1932	Studies of trout producing lakes and ponds. University of Toronto Studies, Biological Series, No. 36. (Publications of the Ontario Fisheries Research Laboratory, No. 45): 113–167.
7	5a	1933	Gregory Clark & W.E. Ricker. The speckled trout and its conservation. Ontario Department of Game and Fisheries, Biological and Fish Culture Branch, Bulletin No. 3, 11 pp.
8	6	1932	The utility of nets in freshwater plankton investigations. Transactions of the American Fisheries Society 62: 292–303. [FRB, S.S. No. 507].
9	7	1933	Destruction of sockeye salmon by predatory fishes. Biological Board of Canada, Progress Reports of the Pacific Coast Stations, No. 18: 3–4.
10	8*	1934	An ecological classification of certain Ontario streams. University of Toronto Studies, Biological Series No. 37, pp. 1–114. (Publications of the Ontario Fisheries Research Laboratory, No. 49): 1–114. [FRB, S.S. No. 507].

KR	WER	Date	
11	9	1934	Plankton organisms and their relation to the sockeye of Cultus Lake. Biological Board of Canada, Progress Reports of the Pacific Coast Stations, No. 21: 14–17.
12	10*	1934	A critical discussion of various measures of oxygen saturation in lakes. Ecology, 15(4): 348–363 [FRB, Studies No. 125].
13	A	1935	Statistical treatment of sampling processes useful in the enumeration of plankton organisms. Fisheries Research Board of Canada, Manuscript Reports of the Biological Stations. No. 255: 19 pp.
14	11	1935	[W.E. Ricker and E.B.S. Logier]. Notes on the occurrence of the ribbed toad (*Ascaphus truei* Stejneger) in Canada. Copeia, 1935(1): 46 [FRB, S.S. No. 567].
15	12	1935	[W.E. Ricker & A. Robertson]. Observations on the behaviour of adult sockeye salmon during the spawning migration. Canadian Field-Naturalist, 49(8): 132–134 [FRB, S.S. No. 568].
16	13	1935	Description of three new Canadian perlids. Canadian Entomologist, 67: 197–201 [FRB, S.S. No. 565].
17		1935	Studies of the limnological factors affecting the propagation and survival of the sockeye salmon (*Oncorhynchus nerka*) at Cultus Lake, British Columbia. Ph.D. thesis, University of Toronto. 207 pp.
18	14	1935	New Canadian perlids (Part 2). Canadian Entomologist, 67: 256–264 [FRB, S.S. No. 566].
19.	15.	1937.	Statistical treatment of sampling processes useful in the enumeration of plankton organisms. Archiv für Hydrobiologie, 31: 68–84. [FRB Studies No. 158].
20	16	1937	The concept of confidence or fiducial limits applied to the Poisson frequency distribution. Journal of the American Statistical Association, 32: 349–356 [FRB Studies No. 167].
21	17	1937	Agriculture in our lakes. Western Angler and Hunter, 2(4): 5–7, April 1937. Kamloops, B.C. [FRB, S.S. No. 626].
22	18	1937	Glimpses of fisheries biology and fish culture in Europe. I. The Scandinavian countries. Progressive Fish-Culturist, 31: 29–33 [FRB, S.S. No. 627].
23.	19.	1937.	Glimpses of fisheries biology and fish culture in Europe. II. Germany and France. Progressive Fish-Culturist, 32: 12–15. FRB, S.S. No. 628.
24	20*	1937	Physical and chemical characteristics of Cultus Lake, British Columbia. Journal of the Biological Board of Canada, 3(4): 363–402.
25	21*	1937	The food and the food supply of sockeye salmon (*Oncorhynchus nerka* Walbaum) in Cultus Lake, British Columbia. Journal of the Biological Board of Canada, 3(5): 450–468.
26	22	1937	Glimpses at fisheries biology and fish culture in Europe. III. Austria, Czechoslovakia, Poland, Russia. Progressive Fish-Culturist, 34: 12–14 [FRB, S.S. No. 629].
27	23	1938	Glimpses at fisheries biology and fish culture in Europe. IV. England and Scotland. Progressive Fish-Culturist, 35: 20–23 [FRB, S.S. No. 661].
28	24*	1938	On adequate quantitative sampling of the pelagic net plankton of a lake. Journal of the Biological Board of Canada, 4(1): 19–32.
29	25*	1938	Seasonal and annual variations in the quantity of pelagic net plankton, Cultus Lake, British Columbia. Journal of the Biological Board of Canada, 4(1): 33–47.
30	26*	1938	A comparison of the seasonal growth rates of young sockeye salmon and young squawfish in Cultus Lake. Biological Board of Canada, Progress Reports of the Pacific Coast Stations, No. 36: 3-5.
31	27	1938	[E. M. Walker & W. E. Ricker]. Notes on Odonata from the vicinity of Cultus Lake, B.C. Canadian Entomologist, 70: 144–151 [FRB, Studies No. 200].
32	28	1938	Notes on specimens of American Plecoptera in European collections. Transactions of the Royal Canadian Institute, 22(1): 129–156 [FRB, S.S. No. 663].
33	29	1938	A new stonefly from Baffin Land (Plecoptera, Capniidae). Canadian Entomologist, 70: 173–174 [FRB, S.S. No. 662].
34	30*	1938	"Residual" and kokanee salmon in Cultus Lake. Journal of the Fisheries Research Board of Canada, 4(3): 192–218.
35	31	1938	[R.E. Foerster & W. E. Ricker]. The effectiveness of predator control in decreasing the mortality of young sockeye salmon (*Oncorhynchus nerka* Walbaum). Verhandlungen der Internationalen Vereinigung für theoretische und angewandte Limnologie, 8: 151–167 [FRB, S.S. No. 645].
36	31a	1938	[W.A. Clemens, J.T. Barnaby & W.E. Ricker]. Production in bodies of water. Progressive Fish-Culturist, 39: 36–43 [FRB, S.S. No. 641].
37	32	1939	A preliminary list of stoneflies (Plecoptera) from the vicinity of Cultus Lake, British Columbia. Proceedings of the Entomological Society of British Columbia, 35: 19–23 [FRB Studies No. 208].
38	33	1939	[W.E. Ricker & C.H.D. Clarke]. The birds of the vicinity of Lake Nipissing, Ontario. Contributions of the Royal Ontario Museum of Zoology, No. 16: 25 pp [FRB, S.S. No. 687].

KR	WER	Date	
39	34	1940	On the relation of "catch per unit effort" to abundance and the rate of exploitation. Journal of the Fisheries Research Board of Canada, 5(1): 43–70.
40	35	1940	On the origin of kokanee, a fresh-water type of sockeye salmon. Transactions of the Royal Society of Canada, Series 534 (V): 121-135 [FRB Studies No. 237].
41	35a	1940	[Anon.] Anglers aid scientists by sending fish scales for age determination. Outdoor Indiana, 7(3): 3, 24.
42	35b	1940	[Anon.] Scientists learn many things from reading fish scales sent to them. Outdoor Indiana, 7(4): 11, 28.
43	36*	1941	The consumption of young sockeye salmon by predacious fish. Journal of the Fisheries Research Board of Canada, 5(3): 293–313.
44	37	1940	[W.E. Ricker & John Gottschalk]. An experiment in removing coarse fish from a lake. Transactions of the American Fisheries Society, 70: 382–390.
45	38a	1941	[W.E. Ricker & C.H.D. Clarke]. Red-breasted merganser on Lake Nipissing. Canadian Field-Naturalist, 55: 137–138.
46	38b	1941	[Anon.]. Science points way to better fishing for the anglers of Indiana. Outdoor Indiana, 8(1): 14, 27.
47	38c	1941	[Anon.] Simonton Lake Club adds gravel to marl bottom to make better bluegill fishing. Outdoor Indiana, 8(6): 7.
48	38*	1942	[R.E. Foerster & W.E. Ricker]. The effect of reduction of predaceous fish on survival of young sockeye salmon at Cultus Lake. Journal of the Fisheries Research Board of Canada, 5(4): 315–336.
49	38d	1941	K.F. Lagler & W.E. Ricker. Science aids fishermen with investigation of Footes Lake in Gibson County. Outdoor Indiana, 8(3): 12.
50	38e	1941	M. Runner & W. E. Ricker. Parasites on Indiana fishes are studied and do not make worthless food. Outdoor Indiana, 8(7): 10, 31.
51	39	1942	Purple sandpiper in Indiana. Wilson Bulletin, 54(4): 250.
52	39a	1942	Research on the biology of fishes. Transylvania College Bulletin, 15(7): 81–84.
53	40	1942	[Karl F. Lagler & W.E. Ricker]. Biological fisheries investigations of Foots Pond, Gibson Country, Indiana. Investigations of Indiana Lakes and Streams, 2(3): 47–72.
54	41	1942	[W.E. Ricker & Karl F. Lagler]. The growth of spiny-rayed fishes in Foots Pond, Indiana. Investigations of Indiana Lakes and Streams, 2(5): 85–97.
55	42*	1942	The rate of growth of bluegill sunfish in lakes of northern Indiana. Investigations of Indiana Lakes and Streams, 2(11): 161–214.
56	43*	1942	Creel census, population estimates and rate of exploitation of game fish in Shoe Lake, Indiana. Investigations of Indiana Lakes and Streams, 2(12): 215–243.
57	44	1942	Fish populations of two artificial lakes. Investigations of India.
58	44a	1942	American biological stations. XXIX. Indiana University Biological Station, Winona Lake, Indiana. Turtox News, 20(10): 138–139.
59	44b	1942	[Anon.]. Ton of fish caught from small lake and still its waters are not estimated overfished. Outdoor Indiana, 8(12): 9, 27.
60	44c	1942	[Anon.]. Winter fishing not hurting lakes so far as data have been obtained and compiled. Outdoor Indiana, 8(12): 20.
61	44d	1942	[Anon.]. Lake and stream survey to produce better fishing in Hoosier waters. Outdoor Indiana, 9(8): 8, 16.
62	44e	1942	[Anon.]. Bugs, worms and insects for fish bait turn anglers into entomologists. Outdoor Indiana, 9(5): 11, 15.
63	45*	1943	Stoneflies of southwestern British Columbia. Indiana University Publications, Science Series, No. 12: 145 pp. [FRB, S.S. No. 766].
64	45a	1943	[Anon.]. Lake and stream data to be collected as practical aid to fishery problems. Outdoor Indiana, 10(5): 12.
65	45b	1943	[Anon.]. Release of tagged trout provides valuable data for better fishing. Outdoor Indiana, 10(11): 8, 9, 13.
66	46	1944	A bird quiz. Indiana Audubon Society Year Book, 21 (1943): 23–27.
67	47*	1944	Further notes on fishing mortality and effort. Copeia, 1944(1): 23–44 [FRB, S.S. No. 780].
68	48	1944	Were our lakes overfished? Outdoor Indiana, 11(8): 2, 16.
69	49	1944	Some Plecoptera from the far north. Canadian Entomologist, 76: 174–185.
70	50	1945	[Anon.]. Sport for any fisherman. Outdoor Indiana, 11(12): 2, 16.
71	51	1945	They're called croppies in Indiana. Outdoor Indiana, 12(3): 2, 6.

KR	WER	Date	
72	52	1945	Causes of death among Indiana fishes. Transactions North American Wildlife Conference, 10: 266–269.
73	53	1945	[W.E. Ricker & Daniel Merriman]. On the methods of measuring fish. Copeia, 1945(2): 185–191.
74	54	1945	Natural mortality among Indiana bluegill sunfish. Ecology, 26(2): 111–121.
75	55*	1945	A method of estimating minimum size limits for obtaining maximum yield. Copeia, 1945(1): 84–94. [FRB, S.S. No. 802].
76	56	1945	Some applications of statistical methods to fishery problems. Biometrics Bulletin, 1(6): 73–79 [FRB, S.S. No. 803].
77	57	1945	A first list of Indiana stoneflies. Proceedings of the Indiana Academy of Science, 54: 225–230.
78	58	1945	Fish catches in three Indiana lakes. Investigations of Indiana Lakes and Streams, 2(16): 325–344.
79	59*	1945	Abundance, exploitation and mortality of the fishes in two lakes. Investigations of Indiana Lakes and Streams, 2(17): 345–448.
80	60	1945	Meet the warmouth bass. Outdoor Indiana, 12(7): 5.
81	60a	1945	[Anon.]. Winter kill – fish loss heavy in northern lakes. Outdoor Indiana, 12(4): 7.
82	60b	1945	[Anon.]. For more and better fish. Outdoor Indiana, 12(7): 15.
83	61	1946	[Anon.]. The "croaker" – white perch. Outdoor Indiana, 13(1): 6.
84	62	1946	What makes our game fish die? Outdoor Indiana, 13(3): 2–3.
85	63	1943	[W.E. Ricker, H.F. Mosbaugh & Maurice Lung]. Production of Indiana hatcheries in 1942. Transactions of the American Fisheries Society, 73: 373–376.
86	64	1946	Industrial and domestic wastes in relation to the aquatic life of Indiana streams. Purdue University Engineering Bulletin, 30(2) (Extension Series No. 60): 90–96.
87	65.	1947.	Hell's Gate and the sockeye. Journal of Wildlife Management, 11(1): 10–20. [FRB, S.S. No. 834].
88	66.	1946.	Some prairie stoneflies (Plecoptera). Transactions of the Royal Canadian Institute, 26(1): 3–8.
89	67*	1946	Production and utilization of fish populations. Ecological Monographs, 16: 373–391 [FRB, S.S. No. 819].
90	68	1946	[G.L. Clarke, W.T. Edmondson & W.E. Ricker]. Mathematical formulation of biological productivity. Ecological Monographs, 16: 336–337 [FRB, S.S. No. 810].
91	69	1947	Tri-lakes test proves fallacy of overfishing. Outdoor Indiana, 14(3): 4–5.
92	70	1947	They'll die anyhow. Outdoor Indiana, 14(4): 2–3.
93	70a	1947	Fishing season extension gets wide approval. Outdoor Indiana, 14(5): 5.
94	70b	1947	[Anon.]. Editors: relax, please! American Scientist, 35(3): 306–318.
95	71	1948	[Anon.]. Let's use our club ponds for fishing. Outdoor Indiana, 15(2): 14–15.
96	72	1948	Stoneflies of the maritime provinces and Newfoundland. Transactions of the Royal Canadian Institute, 26(2): 401–414, plate X.
97	73*	1948	[W.E. Ricker & R.E. Foerster]. Computation of fish production. Bulletin of the Bingham Oceanographic Collection, Yale University, 11(4): 173–211 [FRB Studies No. 295].
98	74*	1948	Methods of estimating vital statistics of fish populations. Indiana University Studies, Science Series, No. 15: 101 pp. [FRB, S.S. No. 860].
99	74a	1948	Methods of estimating vital statistics of fish populations. [partial translation into Spanish]; in Boletin de la Compañía Administradora del Guano (Lima), 28(1–3): 1–16.
100	75	1945	Hybrid sunfish for stocking small ponds. Transactions of the American Fisheries Society, 75: 84–96.
101	75a	1948	Questions and answers concerning fishing seasons. Outdoor Indiana, 15(5): 2, 3.
102	75b	1948	Waterweeds – some are helpful; too many, no. Outdoor Indiana, 15(9): 15: 16.
103	75c	1948	Game fish multiply despite fishing in Hoosier streams. Outdoor Indiana, 15(11): 17.
104	76	1949	Effects of removal of fins upon the growth and survival of spiny-rayed fishes. Journal of Wildlife Management, 13(1): 29–40.
105	77a	1949	Our eight-footed friends. Canadian Field-Naturalist, 63(3): 95–98.
106	77b	1949	Our eight-footed friends. Outdoor Indiana, 16(12): 15.
107	78	1949	[Anon.]. 14 years under water. Outdoor Indiana, 16(10): 8, 20.
108	79	1949	[W.E. Ricker, H.F. Mosbaugh & Maurice Lung]. Utilization and survival of trout in Indiana. Investigations of Indiana Lakes and Streams, 3(6): 271–281.
109	80	1949	Utilization of food by bluegills. Investigations of Indiana Lakes and Streams, 3(8): 311–318.
110	81	1948	[W.E. Ricker & L.A. Krumholz]. Pond management for Indiana ponds. Indiana Department of Conservation, 10 pp.
111	82	1949	The North American species of Paragnetina (Plecoptera, Perlidae). Annals of the Entomological Society of America, 42(3): 279–288.

KR	WER	Date	
112	83	1947	Mortality rates in some little-exploited populations of freshwater fishes. Transactions of the American Fisheries Society, 77: 114–128.
113	84*	1950	Cycle dominance among the Fraser sockeye. Ecology, 31(1): 6–26 [FRB, S.S. No. 919].
114	85	1949	[Anon.]. How much does a fish eat? Outdoor Indiana, 16(4): 2, 19.
115	86	1949	[Anon.]. Minnows most appetizing live bait: say crappies. Outdoor Indiana, 16(7): 16.
116	87*	1950	Some evolutionary trends in Plecoptera. Proceedings of the Indiana Academy of Science, 59: 197–209.
117	87a	1950	[Anon. – prepared by W.E. Ricker & Howard H. Michaud]. Conservation of fishes in Indiana. Unit V of Conservation Education, Indiana Department of Public Instruction, Bulletin No. 173–175: 52 pp.
118	88*	1952	Systematic studies in Plecoptera. Indiana University. Publications, Science Series, No. 18: 200 pp [FRB, S.S. No. 973].
119	89	1952	The benthos of Cultus Lake. Journal of the Fisheries Research Board of Canada, 9(4): 204–212.
120	90	1952	Numerical relations between abundance of predators and survival of prey. Canadian Fish Culturist, No. 13:. 5–9. [FRB Studies No. 340].
121	90a	1952	[Translated title to Japanese by H. Kashara]. Production and utilization of fish populations. Fisheries Agency (Japan), Research Division, Fisheries Science Series No.2: 9–34 pp.
122	B*	1953	[R.E. Foerster & W.E. Ricker]. Predator control and sockeye salmon production. Fish. Res. Bd. Can., Manuscript Reports of the Biological Stations. No. 478: 18 pp.
123	91*	1953.	[R.E. Foerster and W.E. Ricker]. The coho salmon of Cultus Lake and Sweltzer Creek. Journal of the Fisheries Research Board of Canada, 10(6): 293–319.
124	92	1953	Effects of protective and regulatory legislation upon the Pacific herring and halibut fisheries. Canadian Fish Culturist, No. 14: 13–19 [FRB Studies No. 364].
125	92a	1953	[Translated title to Japanese by M. Kono]. Evolutionary trands in Plecoptera. Magazine (?) of Insects, 1953(12): 25 pp.
126	C	1954	[W.E. Ricker, Dorothy Manzer & Elizabeth Neave]. The Fraser River eulachon fishery, 1941–1953. Fish. Res. Bd. Can., Manuscript Reports of the Biological Stations. No. 583: 35 pp.
127	93	1954	Effects of compensatory mortality upon population abundance. Journal of Wildlife Management, 18(1): 45–51 [FRB Studies No. 359].
128	94*	1954	Stock and recruitment. Journal of the Fisheries Research Board of Canada, 11(5): 559–623.
129	94a*	1957	Zapas i popolnenie, pp. 51–93, in Materialy Mezhdunarodnoi Konferentsii po Okhrane Zapasov Ryb Vol. 1 [Moscow, 1957].
130	94b*	1983	Stock and recruitment, pp. 151–225; in Cushing, D.H. (editor) Key Papers in Fish Populations. IRL Press, Oxford-Washington, 405 pp.
131	95	1954	Review: "Haddock bionomics. I. The state of the haddock stocks in the North Sea 1946–50 and at Faroe 1914–50", by B.B. Parrish and R. Jones. Journal du Conseil pour l'Exploration de la Mer, 19(3): 386–387.
132	96	1954	Pacific salmon for Atlantic waters? Canadian Fish Culturist, 16: 6–14
133	96a	1955	Pacific salmon for Atlantic waters? [Norwegian translation by Sven Somme]. Jeger og Fisker (Oslo), 1955, No. 5: 241–248, [No. 6: 269–272.] [FRB Studies No. 381].
134	97	1954	Nomenclatorial notes on Plecoptera. Proceedings of the Entomological Society of British Columbia, 51: 37–39 [FRB, S.S. No. 1046].
135	97a	1954	Reviews: "Commercial fishes of the USSR" (ed. by L.S. Berg et al.) Fed. Fish. Res. Inst. for Fisheries Oceanography. Pischchepromizdat. Moscow. 1949. Vol.1 and Vol.2, 787, and 230 Plates.
136	98	1955	[W.E. Ricker and W.W. Judd]. A collection of stoneflies (Plecoptera) from the Northwest Territories, Canada. Entomological News, 66(10): 255–256 [FRB, S.S. No. 1090].
137	99*	1955	Fish and fishing in Spear Lake, Indiana. Investigations of Indiana Lakes and Streams, 4(5): 117–161 [FRB, S.S. No. 1089].
138/139	100	1955	Reviews: "The natural regulation of animal numbers", by David Lack; and of "The distribution and abundance of animals", by H.G. Andrewartha and L.C. Birch. Journal of Wildlife Management, 19(4): 487–489.
140	101	1956	The marking of fish. Ecology, 37(4): 665–670 [FRB Studies No. 443].
141	101a	1957	[Yvonne Bishop, N.M. Carter, Dorothy Gailus, W.E. Ricker & J.M. Speirs]. Index and list of titles, publications of the Fisheries Research Board of Canada, 1901–1954. Fisheries Research Board of Canada, Bulletin 110: 209 pp.
142	101b		[Y. Bishop, N. M. Carter, W.E. Ricker and D. Gailus] Index to Manuscript Reports of the Biological Stations No. 1–600, and Manuscript Reports of the Technological stations, No. 1–57; together with the titles of the papers in these two series. Fisheries Research Board of Canada. MS Report.

KR	WER	Date	
143	102	1958	Production, reproduction and yield. Verhandlungen der Internationalen Vereinigung für theoretische und angewandte Limnologie, 13: 84–100 [FRB Studies No. 484].
144	103	1958	Some principles involved in regulation of fisheries by quota. Canadian Fish Culturist, No. 22: 1–6 [FRB Studies No. 535].
145	103a	1958	The quota system of regulation: how it works, its advantages and drawbacks. Western Fisheries, 56(4): 10–11, 45–47, 49–50.
146	104*	1958	Maximum sustained yields from fluctuating environments and mixed stocks. Journal of the Fisheries Research Board of Canada, 15(5): 991–1006.
147	105	1958	Handbook of computations for biological statistics of fish populations. Fisheries Research Board of Canada, Bulletin 119: 300 pp.
148	105a	1958	[N.M. Carter & W.E. Ricker, (eds.)]. Index and list of titles for the Circulars and Industrial Memoranda issued by the Stations and Units of the Fisheries Research Board of Canada, 1935–1957. Fisheries Research Board of Canada, Biological Station, Nanaimo, B.C., Circular No. 47: 50 pp.
149	D.	1959.	[Editor]. Survey of Canadian Fisheries. Fish. Res. Bd. Can., Manuscript Reports of the Biological Stations. No. 675: 280 pp.
150	106	1959	The species of *Isocapnia* Banks (Insecta, Plecoptera, Nemouridae). Canadian Journal of Zoology, 37(5): 639–653 [FRB Studies No. 843].
151	107	1959	Additional observations concerning residual sockeye and kokanee(*Oncorhynchus nerka*). Journal of the Fisheries Research Board of Canada, 16(6): 897–902.
152	108	1959	Plecoptera. Chapter 35, pp. 941–957: in W.T. Edmondson (editor) Ward and Whipple's Freshwater Biology, 2nd edition. John Wiley, New York. [FRB, S.S. No. 1229].
153	E	1960	Evidence for environmental and genetic influence on certain characters which distinguish stocks of the Pacific salmons and steelhead trout. Fish. Res. Bd. Can., Manuscript Reports of the Biological Stations. No. 695: 103 pp, 15 tables.
154	109	1960	A population of dwarf coastrange sculpins (*Cottus aleuticus*). Journal of the Fisheries Research Board of Canada, 17(6): 929–932.
155	110	1960	Vliyanie faktorov sredy i nasledstvennosti na morfologiyu i povedenie tikhookeanskogo lososya Severnoi Ameriki. [The effect of environmental factors and heredity on the morphology and behaviour of the Pacific salmon of North America.] VNIRO, Nauchnaya Informatsiya, 1960 (4): 12 pp., Moscow [translated into Russian by L.S. Bogdanov].
156	111	1960	Review: "Ciba Foundation Colloquia on Ageing. Volume 5. The lifespan of animals." Journal du Conseil pour l'Exploration de la Mer, 26 (1): 125–129.
157	112	1961	Plecoptera. pp. 810–811; in "The Encyclopedia of the Biological Sciences." Reinhold Publishing Corporation, New York, 1119 pp. [FRB, S.S. No. 1303].
158	113	1961	Kratkaya istoriya razvitiya teorii dinamiki chislennosti populyatsii ryb v stranakh zapadnoi Evropy I Severnoi Ameriki. [A short history of the development of the theory of the dynamics of abundance of fish populations in the countries of western Europe and North America], pp. 211–215 in Trudy Soveshchanii Ikhtiologicheskoi Komissii Akademii Nauk SSSR, No. 13: 1961. Soveshchanie po dinamike chislennosti ryb, 1960. Translated into Russian by L.S. Bogdanov. [FRB, S.S. No. 1302].
159	114(IV)	1961	75th birthday of Professor F.I. Baranov. Copeia, 1961(4): 517–518.
160	114a*	1961	Review: "Pêches continentales – biologie et amènagement", by R. Vibert & K.F. Lagler. Canadian Fish Culturist, 29: 41–42.
161	114b	1961	Review: "Pêches continentales – biologie et amènagement", by R. Vibert & K.F. Lagler. Copeia for 1962 (No. 1): 242–243.
162	115a*	1961	Productive capacity of Canadian fisheries – an outline, pp. 775–781. Background papers, "Resources for Tomorrow" Conference, Vol. 2. Ottawa.
163	115c*		Productive capacity of Canadian fisheries: an outline. Trade News, 14(2): 5–7 [FRB Studies No. 1304].
164	116	1962	Connecticut River shad. Transactions of the American Fisheries Society, 91(1): 98–99 [FRB, S.S. No. 1341].
165	117	1962	[W.E. Ricker & Ferris Neave]. Nesting colony of mew gulls on Kennedy Lake, Vancouver Island. Report of the British Columbia Provincial Museum of Natural History and Anthropology 1961: 20–21 [FRB, S.S. No. 1342].
166	118	1962	Review: "Experimental ecology of the feeding of fishes", by V.S. Ivlev. Transactions of the American Fisheries Society, 91(2): 237–238

KR	WER	Date	
167	119*	1962	Regulation of the abundance of pink salmon populations. pp. 125–201; in N.J. Wilimovsky (ed.), Symposium on Pink Salmon, University of British Columbia, H.R. McMillan Lectures in Fisheries, 1960, Vancouver, B.C. [FRB Studies No. 713].
168	120*	1962	Comparison of ocean growth and mortality of sockeye salmon during their last 2 years. Journal of the Fisheries Research Board of Canada, 19(4): 531–560.
169	121	1962	Fish culture. Vol. 9, pp. 291–291F in Encyclopaedia Britannica, 1962 edition. [FRB, S.S. No. 1339].
170	122	1962	Plecoptera. Vol. 18, pp. 72–73 in Encyclopaedia Britannica, 1962 edition [FRB, S.S. No. 1340].
171	123	1962	[Editor]. Productive capacity of Canadian fisheries. Fisheries Research Board of Canada, Nanaimo Biological Station, Circular No. 64: 79 pp.
172	124	1962	Russian-English glossary of names of aquatic organisms and other biological and related terms. Fisheries Research Board of Canada, Nanaimo Biological Station, Circular No. 65: 173 pp.
173	F	1963	Distribution of Canadian stoneflies. Fish. Res. Bd. Can., Manuscript Reports of the Biological Stations. No. 770: 24 pp, 47 maps.
174	125	1963	Big effects from small causes: two examples from fish population dynamics. Journal of the Fisheries Research Board of Canada, 20(2): 257–264.
175	125a	1963	[Statistical appendix to paper by D.B. Quayle.] Mortality in Pacific oyster seed. Proceedings of the National Shellfisheries Association, 52: 53–63, [FRB Studies No. 813].
176	126	1964	[P.A. Larkin & W.E. Ricker, (eds.)]. Canada's Pacific marine fisheries: past performance and future prospects, pp. 194–268; in Inventory of the Natural Resources of British Columbia, 1964, Published by the British Columbia National Resources Conference. Victoria, B.C, 610 pp. [FRB Studies No. 851].
177	127	1964	[H.H. Ross & W.E. Ricker]. New species of winter stoneflies, genus *Allocapnia* (Plecoptera, Capniidae). Transactions of the Illinois Academy of Science, 57(2): 88–93 [FRB Studies No. 872].
178	128	1964	[P.A. Larkin & W.E. Ricker]. Further information on sustained yields from fluctuating environments. Journal of the Fisheries Research Board of Canada, 21(1): 1–7.
179	129	1964	The need for habitat. pp. 29–38; in J.R. Dymond (ed.) "Fish and Wildlife": A Memorial to W.J.K. Harkness. Longmans Canada Ltd. Toronto. 214 pp. [FRB, S.S. No. 1397].
180	130	1964	[G.F.M. Smith & W.E. Ricker]. Availability of fish. Canadian Fisheries Reports, 3: 25–29 [FRB Studies No. 900].
181	131	1964	Ocean growth and mortality of pink and chum salmon. Journal of the Fisheries Research Board of Canada, 21(5): 905–931.
182	132*	1969	Distribution of Canadian stoneflies. Gewasser und Abwasser, Heft 34/35: 50–71.
183	133	1965	New records and descriptions of Plecoptera (Insecta). Journal of the Fisheries Research Board of Canada, 22(2): 475–501.
184	134	1965	[H.T. Bilton & W.E. Ricker]. Supplementary checks on the scales of pink salmon (*Oncorhynchus gorbuscha*) and chum salmon (*O. keta*). Journal of the Fisheries Research Board of Canada, 22(6): 1477–1489.
185	135	1966	Salmon of the North Pacific Ocean – Part III. A review of the life history of North Pacific salmon. 4. Sockeye salmon in British Columbia. Bulletin of the International North Pacific Fisheries Commission, 18: 59–70 [FRB Studies No. 1062].
186	G	1967	[W.E. Ricker & J.I. Manzer]. Recent information on salmon stocks in British Columbia. Fish. Res. Bd. Can., Manuscript Reports of the Biological Stations. No. 931: 45 pp., 13 tables and 18 figures.
187	136	1968	[W.E. Ricker & K.H. Loftus]. Pacific salmon move east. Fisheries Council of Canada, Annual Review, 1968: 37–39, 43 [FRB No. A106].
188	H	1968	Background information and theory related to management of the Skeena sockeye salmon. Fish. Res. Bd. Can., Manuscript Reports of the Biological Stations. 961, 38 pp.
189	137	1968	Introduction (Chapter 1), Appendices, and Index. pp. 1–6, 255–269 and 307–313; in W.E. Ricker (ed.) "Methods for assessment of fish production in fresh waters". International Biological Program Handbook No. 3: 313 pp. [2nd edition. Published in 1971].
190	138*	1968	[W.E. Ricker & H.H. Ross]. North American species of *Taeniopteryx* (Plecoptera, Insecta). Journal of the Fisheries Research Board of Canada, 25(7): 1423–1439
191	139	1968	[Editor]. A short dictionary of fishery terms (Russian–English). Compiled by A.A. Klykov. Translated and published by U.S. Department of Commerce, Clearing House for Federal Scientific and Technical Information, Joint Publications Research Service. 67 pp.
192	139a	1968	[Anon.]. The Fisheries Research Board of Canada – Planning for the next decade, 1968–1978. Fisheries Research Board of Canada, Annual Meeting, January 1968; 51 pp.

KR	WER	Date	
193	140	1968	[W.E. Ricker, Renè Malouin, Peter Harper & H.H. Ross]. Distribution of Quèbec stoneflies (Plecoptera). Le Naturaliste Canadien, 95: 1085–1123. [FRB Studies No. 1268].
194	141	1968	[L.V. Pienaar & W.E. Ricker]. Estimating mean weight from length statistics. Journal of the Fisheries Research Board of Canada, 25(12): 2743–2747.
195	142*	1969	Effects of size-selective mortality and sampling bias on estimates of growth, mortality, production, and yield. Journal of the Fisheries Research Board of Canada, 26(3): 479–541.
196	143	1969	Review: "Fisheries biology. A study in population dynamics", by D.H. Cushing. Transactions of the American Fisheries Society, 98(2): 341–342.
197	144*	1969	[W.E. Ricker & H.H. Ross]. The genus *Zealeuctra* and its position in the family Leuctridae (Plecoptera, Insecta). Canadian Journal of Zoology, 47(6): 1113–1127 [FRB Studies No. 1375].
198	145*	1969	Food from the sea. Chapter 5 (pp. 87–108); in Preston Cloud (chair) Resources and man, a study and recommendations. Report of the Committee on Resources and Man, U.S. National Academy of Sciences, W.H. Freeman and Co., San Francisco, 259 pp.
199	K	1969	[P.A. Larkin, J. McDonald, R.R. Parker, F. Neave, H. Godfrey & W.E. Ricker]. Research programs concerned with methods of increasing salmon populations. Fish. Res. Bd. Can., Manuscript Reports of the Biological Stations. No. 1015: 56 pp.
200	146	1969	[Arthur Tautz, P.A. Larkin & W.E. Ricker]. Some effects of simulated long-term environmental fluctuations on maximum sustained yield. Journal of the Fisheries Research Board of Canada, 26(10): 2715–2726.
201	I	1970	Destruction of young salmon during their seaward migration. Fish. Res. Bd. Can., Manuscript Reports of the Biological Stations. No. 1081: 12 pp.
202	147	1970	Biostatisticheskii metod A.N. Derzhavina. [A.N. Derzhavinís biostatistical method]. Translated by A.V. Zasizov. Rybnoe Khozyaistvo, 46 (10): 6–9; 46(11): 5–7 [FRB Studies No. 1582].
203	J	1970	Some fishery research stations in the USSR, 1969. Fish. Res. Bd. Can., Manuscript Reports of the Biological Stations. No. 1091: 109 pp.
204	148	1971	Sopostavlenie dvukh krivykh vosproizvodstva. [Comparison of two reproduction curves]. Translated by A.V. Zasizov. Rybnoe Khozyaistvo, 47(3): 16–21; 47(4): 10–13. [FRB Studies No. 1583].
205	L	1971	Lists of active research projects related to the aquatic environment and its inhabitants, conducted by various organizations in the Pacific region. Fish. Res. Bd. Can., Manuscript Reports of the Biological Stations. No. 1139: 60 pp.
206	149	1971	Review: "Biologicheskie resursy Mirovogo okeana", by P.A. Moiseev. Transactions of the American Fisheries Society, 100(3): 593–594
207	150	1970	[P.A. Larkin, D. Pimlott & W.E. Ricker]. Vertebrate ecology: a review of present status. Report No. 5, 40 pp; in Panel Reports of the Study of Basic Biology in Canada: by the Biological Council of Canada and the Canadian Federation of Biological Sciences. Biological Council of Canada, Ottawa.
208	151*	1971	[H.H. Ross & W.E. Ricker]. The classification, evolution and dispersal of the winter stonefly genus *Allocapnia*. Illinois Biological Monographs, No. 45: 166 pp. [FRB Studies No. 1647].
209	152	1971	Derzhavinís biostatistical method of population analysis. Journal of the Fisheries Research Board of Canada, 28 (10): 1666–1672.
		1972	[Ross, H.H. & W.E. Ricker]. The classification, evolution and dispersal of the winter stonefly genus *Allocapnia*. J. Kansas Entomological Society, 45(1): 141–143 (Review of 208 above).
210	153	1972	Research and management. Transactions of the American Fisheries Society, 101(1): 150–157 [FRB, S.S. No. 1756].
211	154*	1972	Hereditary and environmental factors affecting certain salmonid populations. pp. 27–160; in The Stock Concept in Pacific Salmon. University of British Columbia, H.R. McMillan Lectures in Fisheries, 231 pp. Vancouver, B.C. [FRB A 259].
212	155	1972	[A.R. Gaufin, W.E. Ricker, Michael Miner, Paul Milam & R.A. Hays]. The stoneflies (Plecoptera) of Montana. Transactions of the American Entomological Society, 98: 161 pp. [FRB Studies No. 1764].
213	156*	1973	Linear regressions in fishery research. Journal of the Fisheries Research Board Canada, 30: 409–434.
214	157*	1973	Two mechanisms that make it impossible to maintain peak-period yields from stocks of Pacific salmon and other fishes. Journal of the Fisheries Research Board Canada, 30(9): 1275–1286.
215	157a	1970	[W.E. Ricker & others]. Report of the Panel of Experts on the population dynamics of Peruvian anchoveta. Boletin del Instituto del Mar del Peru, 2(6): 323–372.
216	158	1972	[W.E. Ricker & others]. Report of the 2nd session of the Panel of Experts on the population dynamics of the Peruvian anchovy. Boletin del Instituto del Mar del Peru, 2(7): 377–457.

KR	WER	Date	
217	159	1973	[W.E. Ricker & others] Third session of the Panel of Experts on the population dynamics of Peruvian anchovy. Boletin del Instituto del Mar del Peru, 2(9): 525–599.
218	160	1973	Critical statistics from two reproduction curves. Rapports et Procès-Verbaux des Réunions du Conseil Permanent International pour l'Exploration de la Mer, 174: 333–340.
219	161*	1973	Russian-English dictionary for students of fisheries and aquatic biology. Fisheries Research Board of Canada Bulletin 183: 428 pp.
220	162	1974	[W.F. Royce, R.C. Croker, D.L. McKernan, W.R. Martin, W.H. Noerenberg, W.E. Ricker & P.M. Roedel]. North American fisheries policy. American Fisheries Society, Washington, D.C. 2 pp.
221	163	1974	[G.I. Murphy, L.K. Boerema, J.A. Gulland, W.E. Ricker, B.J. Rothschild, G.S. Saetersdal, & A. Saville]. Report of the Fourth Session of the Panel of Experts on stock assessment of Peruvian anchoveta. Boletin del Instituto del Mar del Peru, 2(10): 605–723.
222	164	1974	[W.E. Ricker & J.I. Manzer]. Recent information on salmon stocks in British Columbia. Bulletin of the International North Pacific Fisheries Commission, 29: 1–36.
223	164a	1974	[Under pseudonym Sam On]. Historiarum historia (a history of histories). Pacific Biological Station Bulletin, 1(4): 8–9.
224	164b	1974	Reviews: "Specific Production of aquatic invertebrates" by V.E. Zaiko. Wiley, New York and Toronto, 1973: 154 pp. Aquaculture, 4 (1974): 313–314.
225	165*	1975	[W.E. Ricker and H.H. Ross]. Synopsis of the Brachypterinae (Insecta, Plecoptera, Taeniopterygidae). Canadian Journal of Zoology, 53(2): 132–153.
226	166	1975	Review: "Hydrographic history and relict fishes of the North-central Great Basin", by C.L. Hubbs, R.R. Miller and L.C. Hubbs. Transactions of the American Fisheries Society, 104(1): 169.
227	167a*	1975	Computation and interpretation of biological statistics of fish populations. Bulletin of the Fisheries Research Board of Canada, 191: 382 pp.
228	167b*	1980	Calcul et intreprétation des statistiques biologiques des populations de poissons. Bulletin of the Fisheries Research Board of Canada, 191 F: 409 pp.
229	168*	1975	[W.E. Ricker & H.D. Smith]. A revised interpretation of the history of the Skeena River sockeye salmon. Journal of the Fisheries Research Board of Canada, 32: 1369–1381.
230	169a	1975	The Fisheries Research Board of Canada – seventy-five years of achievements. Journal of the Fisheries Research Board of Canada, 32: 1465–1490.
231	169b	1975	L'office des recherches sur les pêcheries du Canada – soixante-quinze années de réalisations. Rapports de l'office des recherches sur les pêcheries du Canada, 2: 26,28 pp.
232	170	1974	[A.R. Gaufin & W.E. Ricker]. Additions and corrections to a list of Montana stoneflies. Entomological News, 85: 285–288.
233	171	1975	[W.E. Ricker & G.G. Scudder]. An annotated checklist of the Plecoptera (Insecta) of British Columbia. Syesis, 8: 285–288.
234	172	1975	A note concerning Professor Jolicoeur's comments. Journal of the Fisheries Research Board of Canada, 32: 1494–1498.
235	172a	1975	Mortality and production of harp seals, with reference to a paper of Benjaminsen and Oritsland (1975). International Commission for the Northwest Atlantic Fisheries, Research Document 75/XII/143, Serial No. 3716.
236	173*	1976	Review of the rate of growth and mortality of Pacific salmon in salt water, and noncatch mortality caused by fishing. Journal of the Fisheries Research Board of Canada, 33: 1483–1524.
237	174	1977	The historical development. pp. 1–26; in "Fish Population Dynamics", J.A. Gulland (ed.). John Wiley and Sons, London. 372 p.
238	174a	1977	Editor of: K. Johnstoneís "The aquatic explorers. A history of the Fisheries Research Board of Canada". University of Toronto Press, Toronto. 342 pp.
239	174b	1977	A landmark demolished. Pacific Biological Station Bulletin, 4(1): 11 pp and photo plates.
240	175	1976	Origin of stonefly names proposed by Ricker and collaborators. Perla, 2: 5–19.
241	175a	1992	Origin of stonefly names proposed by Ricker and collaborators. Perla 2: 5–19 (reprinted by Perla, vol. 18, with additions and emendations, 1992.)
242	176	1978	[S.J. Westrheim and W.E. Ricker]. Bias in using an age-length key to estimate age-frequency distributions. Journal of the Fisheries Research Board of Canada, 35(2): 184–189.
243	177	1978	On computing production. Limnology and Oceanography, 23(2): 379–380.
244	178	1978	[W.E. Ricker, H.T. Bilton & K.V. Aro]. Causes of the decrease in size of pink salmon (*Oncorhynchus gorbuscha*). Fisheries and Marine Service Technical Report No. 820: 93 pp.

KR	WER	Date	
245	178a	1979	Ricker, Y.E. [Kryllish script used] Metod oshenki i interpretashiia biologicheskikh pokazatelei populiashii ryb. Pisheheraji Prormishlennosti, Moscow, 408 pp.
246	179	1979	Growth rates and models. Chapter 11, pp. 667–743; in W.S. Hoar, D.J. Randall and J.R. Brett (eds.), "Bioenergetics and Growth". Fish Physiology, vol. 8. Academic Press, New York.
247	179a	1979	Obituary: Neal Marshall Carter. Journal of the Fisheries Research Board of Canada, 36: 595–596.
248	179b	1979	Obituary: Russell Earle Foerster. Journal of the Fisheries Research Board of Canada, 36: 1292–1293.
249	180	1980	Changes in the age and size of chum salmon (*Oncorhynchus keta*). Canadian Technical Report of Fisheries and Aquatic Sciences, 930: 99 p.
250	181	1980	Causes of the decrease in age and size of chinook salmon (Oncorhynchus tshawytscha). Canadian Technical Report of Fisheries and Aquatic Sciences, 944: 25 pp.
251	182	1980	[W.E. Ricker & W.P. Wickett]. Causes of the decrease in size of coho salmon (Oncorhynchus kisutch). Canadian Technical Report of Fisheries and Aquatic Sciences, No. 971: 63 pp.
252	183*	1981	Changes in the average size and average age of Pacific salmon. Canadian Journal of Fisheries and Aquatic Sciences, 38: 1636–1656.
253	M	1982	[W.E. Vanstone & W.E. Ricker]. The IDRC milkfish project, 1975–1978. International Development Research Centre, MS Report 53e (new series): 269 p.
254	184	1982.	This week's citation classic: W.E. Ricker's "Stock and recruitment". Current Contents: Life Sciences, 1982 (21): 20.
255	184a	1982	Obituary: Russell Earle Foerster. Proceedings of Royal Society of Canadian Series IV, 20: 85–86.
256	185	1982	Size and age of British Columbia sockeye salmon (*Oncorhynchus nerka*) in relation to environmental factors and the fishery. Canadian Technical Report of Fisheries and Aquatic Sciences, 1115: 117 pp.
257	186a	1982.	Linear regressions for naturally variable data. Biometrics, 38(3): 859–860
258	186a	1983	Chapter 6 Kolichestvennye pokazateli i modeli rostaryb: pp: 346–402 in Bioenergetika i Rost Rybjedibg M. I. Shatunovskii and Yu. Yu. Dgebuarze. Light and Food Industries, Academic Press, Moscow, 407 pp.
259	187*	1984	Computation and uses of central trend lines. Canadian Journal of Zoology, 62: 1897–1905.
260	188	1984	Trends in size of British Columbia salmon, 1975–1982. Document submitted to the annual meeting of the International North Pacific Fisheries Commission, October, 1984. 23 pp.
261	188a	1984	Who was our first fly fisherman? American Fisheries Society, North Pacific International Chapter, Newsletter.
262	188b	1985	Who was our first fly fisherman? Totem Topics, Newsletter of the Totem Flyfishers, 77: 7–8.
263	189	1985	This week's citation classic: W.E. Ricker's "Linear regressions in fishery research". Current Contents: Life Sciences, 1985 (27): 20.
264	190	1985	[Three Haiku]. Seki Jin (Japan), 1985(2): 26.
265	191	1985	Varieties and properties of three recruitment curves. La Mer [Tokyo], 23: 153–157.
266	192	1986	[Karl Ricker & W.E. Ricker]. More monkey wrenching with the plumbing of Ape Lake and its surrounding glaciers. The Canadian Alpine Journal, 69: 45–47.
267	192a	1986	[W.E. Ricker & J.R. Brett]. Obituary: Ferris Neave, 1901–1986. Trans. Royal Soc. Can. Series V, I, 328–331.
268	193*	1987	Effects of the fishery and of obstacles to migration on the abundance of Fraser River sockeye salmon (*Oncorhynchus nerka*). Canadian Technical Report of Fisheries and Aquatic Sciences, 1522: 75 pp.
269	194*	1988	The absent corpse mystery solved; or, Professor Moriarty in the Yukon. Canadian Holmes, 12(2): 3–11.
270/271	194a/b	1997	The absent corpse mystery solved; or, Professor Moriarty in the Yukon. pp. 211–218; in C. Redman, (ed). "Canadian Holmes, the first twenty-five years - an anthology". Calabash Press, Ashcroft, B.C., 304 pp. (reprinted in The Petrel Flyer (Newsletter of the Stormy Petrels of B.C.; vol 12(1):21–23, 12(2):24–26, 12(3):in press, in 3 parts, 2000)
272	195	1989	History and present state of the odd-year pink salmon runs of the Fraser River region. Canadian Technical Report of Fisheries and Aquatic Sciences, 1702: 37 pp.
273	196	1990	One hundred years of computations. (re – W.E. Ricker's Computation and Interpretation of Biological Statistics of Fish Populations). Current Contents: Life Sciences, 1990 (21–39): 18.
274	197	1991	Meteors over B.C. in 1913. British Columbia Historical News, Spring 1991, p. 32.
275	197a	1991	Inductive or deductive? The Petrel Flyer (Newsletter of the Stormy Petrels of B.C.), 3(2): 1, 8.
276	198*	1992	Back-calculation of fish lengths based on proportionality between scale and length increments. Canadian Journal of Fisheries and Aquatic Sciences, 49(5): 1018–1026.

KR	WER	Date	
277	199	1993	The name is the message. American Fisheries Society, Membership Directory and Handbook, 1992–93: 189–190.
278	199a	1994	The absent corpse mystery solved; or, Professor Moriarity in the Yukon; 108–118, in "The men with the twisted kongo", ed. by H. Yuichi and K.W. Webb. The Shoso-in Bulletin, Tokyo, 4: 130 pp.
279	200	1994	Fowl play. The Petrel Flyer (Newsletter of the Stormy Petrels of B.C.), 6 (8): 5, 7.
280	200a	1995	Fowl play, pp 30–39 in M. Hughes, K.B. Webb and H. Vuichi (eds.), "The Men with the Twisted Kenjo". The Shos-in Bull. Vol. 5 (International Series): 152 pp.
281	201	1994	[P. Harper, & W.E. Ricker]. Distribution of Ontario stoneflies (Plecoptera). Proceedings Entomological Society of Ontario, 125: 43–66.
282	202	1995	Trends in the average size of Pacific salmon in Canadian catches. pp. 593–602; in R.J. Beamish, (ed.) "Climate change and northern fish populations". Canadian Special Publication of Fisheries and Aquatic Sciences No. 121: 739 pp.
283	203	1995	Moriarity, Fermat & Einstein: an addendum to "The Absent Corpse Mystery Solved". Canadian Holmes, 19: 10–11.
284	204*	1996	Stonefly names from classical times. Perla, 14: 37–43.
285	205*	1997	Cycles of abundance among Fraser River sockeye (*Oncorhynchus nerka*). Canadian Journal of Fisheries and Aquatic Sciences, 54: 950–968.
286	206*	1997	[K.W. Stewart, & W.E. Ricker]. Stoneflies (Plecoptera) of the Yukon; Chapter 11, pp. 201–222; in H.V. Danks and J.A. Downes (Eds.), Insects of the Yukon. Biol. Surv. of Canada (Terrestrial Arthropods), Ottawa, 1034 pp.
287	207a	1995	Book Review: R.C. Weyman, "Sherlock Holmes in the Canadian West". 1994. Simmon and Pierre, Toronto, 136 pp. Canadian Holmes, 20 pp.
288	207b	1995	Book Review: R.C. Weyman, "Sherlock Holmes in the Canadian West". 1994. The Petrel Flyer (Newsletter, Stormy Petrels of B.C.), 9(2): 4–5.
289	208	1997	That Motto. pp. 236; in C. Redman, (ed.), Canadian Holmes, the first twenty-five years–an anthology. Calabash Press, Ashcroft, B.C. 304 pp.
290	209	1997	Letter re Arthur C. Clarke description of Conan Doyle as a "raving lunatic". Petrel Flyer (Newsletter, Stormy Petrels of B.C.), 9(4): p. 2.
291	210	1999	[Ricker, W.E. & J.T. Schnute.]. A westward trajectory extension for the celestial fireballs seen in 1913. Canadian Field-Naturalist, 113: 693–697.
292	211	2005	One man's journey through the years when ecology came of age. Environmental Biology of Fishes 74: 000–000.
293	212	1998	Letter: re-Bill Hill's letter, and H.M.S. Bounty. The Petrel Flyer (Newsletter, Stormy Petrels of B.C.), 10(5): p. 5.
294	213	1999	Lidija Andrejevna Zhiltsova; pp. 14–15 in "Lifetime Achievement Awards". Internat. Soc. Plecopterists. Perla, No. 17.
295	214	1999	Petrel Profile Member #17, Bill Ricker. The Petrel Flyer (Newsletter of the Stormy Petrels of B.C.) 11(5): 12–14.

*12 additional manuscripts have been published since the compilation of the original manuscript.

Selected published translations (Russian to English)

KR	WER	DATE	
1		1946	On the question of the biological basis of fisheries. By T.I. Baranoff. from Nauchn. Issled. Ikhtiologicheski Inst. Izvestia, 1: 81–128 (1918). [Translated by W.E. Ricker & N. Artin]. Indiana University, Dept. Biol. Sci. 77 pp.
2		1946	On the question of the dynamics of the fishing industry. By T.I. Baranoff. Byull-Rybn. Khoz. (1925) 8: 7–11. (1926). [Translated by W.E. Ricker & N. Artin]. Indiana Univ., Dept. Biol. Sci. 14 pp.
3		n.d.	The capture of fish by gill nets. Chapter 8, in "Tekhnika Promyshlennogi Rybolovsta". By F.I. Baranov. from Mater. Posnaniyu Russ. Rybolov. 3(6): 56–99 (1914) Pacific Biol. Stn. (Nanaimo, B.C.) Reprints, 24 pp.

KR	WER	DATE	
8		1952	The stock of vobla in the North Caspian Sea, and methods of assessing it. By G.H. Monastrysky. from Trudy Vsesoiusnovo N.-I. Instituta Rybnovo Khoziaistva i Okeanografii, Vol. 11, pp. 115–170 (1940) [Translated by W.E. Ricker]. Fish. Res. Bd. Can., Translation Series No. 81: 58 pp.
9		1953	The reaction of young sturgeons to oxygen deficiency in relation to temperature. By A.B. Lozinov. from Zoologicheskii Zhurnal, 31(5): 686–695 [Translated by W.E. Ricker]. Fish. Res. Bd. Can., Translation Series No. 78: 20 pp.
10		1955	Food and growth of young chum salmon in fresh water. By V. Ya. Levanidov. from Zoologicheskii Zhurnal, 34(2): 371–379 (1955) [Translated by W.E. Ricker & Deena Wakhroucheff]. Fish. Res. Bd. Can., Translation Series No. 77: 10 pp.
11		1956	On the biology of reproduction of the White Sea wolf-fish (*Anarhichas lupus L.*). By V.V. Barsukov. from Zoologicheskii Zhurnal, 32(6): 1211–1216 (1953). [Preliminary translation by W.E. Ricker]. Fish. Res. Bd. Can., Translation Series No. 62: 6, 6 pp.
12		1956	Concerning the biological basis of the rate of exploration, and means of managing the abundance of fish stocks. By G.V. Nikolsky. from Ocherki po Obshchim Voprosam Ikhtiologii [Essays on General Problems of Ichthyology], 1953, pp. 306–318. Academy of Sciences of the USSR, Leningrad. [Preliminary translation by W.E. Ricker]. Fish. Res. Bd. Can., Translation Series No. 83: 16 pp.
14		1956	Contribution to a biological foundation for the salmon fishing industry of the Amur basin. By G.V. Nikolsky. from Trudy Soveshchaniia po Voprosam Lososevovo Khoziaistva Dalnevo Vostoka [Proceedings of a Symposium on Problems of the Salmon Industry of the Far East], 1953, pp. 160–168. [Preliminary translation by W.E. Ricker]. Fish. Res. Bd. Can., Translation Series No. 84: 10 pp.
15		n.d.	Some ecologico-physiological characteristics of whitefishes acclimatized in waters of eastern Ukraine. By G.L. Shkorbatov. from Zoologicheskii Zhurnal 33(6): 1325–1335. (1954). [Preliminary translation by W.E. Ricker]. Fish. Res. Bd. Can., Translation Series No. 89: 18 pp.
16		n.d.	A volumetric method of quantitative estimation of young fish. By V.S. Tanasiichuk. from Zoologicheskii Zhurnal, 31(4): 620–633. (1952) [Preliminary translation by W.E. Ricker]. Fish. Res. Bd. Can., Translation Series No. 90: unpaginated.
17		1956	Information on the effect of the fishery on the condition and composition of the stock of Flatfishes in the Gulf of Peter The Great. By P.A. Moiseev. from Zoologicheskii Zhurnal, (5): 451–457 (1946). [Preliminary translation by W.E. Ricker]. Fish. Res. Bd. Can., Translation Series No. 98: 9, 3 pp.
18		1957	Some characteristics of the distribution of bottom and demersal fishes of far-eastern seas. by P.A. Moiseev. from Izvestiia Tihookeanskovo Nauchno-issledovatelskovo Instituta Rybnovo Khoziaistva i Okeanograffi, (37): 129–137 (1952) Vladivostok. [Translated by the Bureau for Translations, Foreign Language Division, Department of the Secretary of State of Canada, edited by W.E. Ricker]. Fish. Res. Bd. Can., Translation Series No. 94: 10 pp.
19		1957	On the dynamics of abundance of the sockeye salmon [*Oncorhynchus nerka* (Walb.)]. By F.V. Krogius. from Izvestiia Tikhookeanskovo Nauchno-issledovatelskovo Instituta Rybnovo Khoziaistva i Okeanograffi, (35): 1–16. (1951). Vladivostok. [Translated by the Bureau for Translations, Foreign Language Division, Department of the Secretary of State of Canada, edited by W.E. Ricker]. Fish. Res. Bd. Can., Translation Series No. 101: 16 pp.
20		1957	The food of young Amur chum salmon in fresh water. By V. Ia. Levanidov and I.M. Levanidova. from Izvestiia Tikhockeanskovo Nauchno-issledovatelskovo Instituta Rybnovo Khoziaistva i Okeanografii, (35): 41–46 (1951) Vladivostok. [Translated by the Bureau for Translations, Foreign Language Division, Department of the Secretary of State of Canada, edited by W.E. Ricker]. Fish. Res. Bd. Can., Translation Series 102: 6 pp.
21		1957	Qualitative characteristics of the stocks and the dynamics of abundance of the autumn chum salmon of the Amur River. By I.B. Birman. from Izvestiia Tikhockeanskovo Nauchno-issledovatelskovo Instituta Rybnovo Khoziaistva i Okeanografii, (35): 17–31 (1951). Vladivostok. [Translated by the Bureau for Translations, Foreign Language Division, Department of the Secretary of State of Canada, edited by W.E. Ricker]. Fish. Res. Bd. Can., Translation Series No. 103: 16 pp.
22		1957	Stocks of Greenland seals and their utilization. By S.V. Dorofeev. from Rybnoe Khoziaistvo, (12): 56–59. (1956). [Preliminary translation by W.E. Ricker]. Fish. Res. Bd. Can., Translation Series No. 113: 6 pp.
23		1957	Biological characteristics of the shore aggregations of the Walrus in the Chukotka Peninsula. By P.G. Nikulin. from Izvestiia Tikhockeanskovo Nauchno-issledovatelskovo Instituta Rybnovo Khoziaistva i Okeanografii, (25): 226–228. (1947). Vladivostok. [Preliminary translation by W.E. Ricker]. Fish. Res. Bd. Can., Translation Series No. 115: 4 pp.

KR	WER	DATE	
24		n.d.	Biology of whales and the whaling fishery in far eastern seas. By M.M. Sleptsov. from Biologiia i promysel kitov dalnevostochnykh morei Fishchepromizdat (Food Industries Press), Moscow. (1955). [Translation of contents and conclusion by W.E. Ricker]. Fish. Res. Bd. Can., Translation Series No. 118.
25		1957	The salmon of south Sakhalin. By P.A. Dvinin. from Izvestiia Tikhockeanskovo Nauchno-issledovatelskovo Instituta Rybnovo Khoziaistva i Okeanografii, (37): 69–108 (1952). Vladivostok. [Translated. By the Bureau for Translations, Foreign Language Division, Department of the Secretary of State of Canada, edited by W.E. Ricker]. Fish. Res. Bd. Can., Translation Series No. 120: 33, 15 pp.
26		1958	Results of investigations of the spawning grounds of the Atlantic-Scandinavian herring. By I.G. Iudanov. from Doklady Akademii Nauk SSSR, 112 (6): 1139–1141. (1957). [Preliminary translation by W.E. Ricker]. Fish. Res. Bd. Can., Translation Series No. 140: 6 pp.
27		1958	Principles of the distribution of fishes and the geological history of the far-eastern seas. By G.U. Lindberg. from Ocherki po Obshchim Voprosam Ikhtiologii, pp. 47–51. (1953). [Akademiia Nauk SSSR, Ikhtiologicheskaia Komissiia, Moscow-Leningrad]. [Preliminary translation by W.E. Ricker]. Fish. Res. Bd. Can., Translation Series No. 141: 12 pp.
28		1958	Concerning the causes of a peculiarity of the pink salmon of the Sea of Japan. By I.B. Birman. from Zoologicheskii Zhurnal, 35 (11): 1681–1684 (1956) [Preliminary translation by W.E. Ricker]. Fish. Res. Bd. Can., Translation Series No. 142: 5 pp.
30		1958	Concerning the specificity of fish parasites. By S.S. Shulman. from Zoologicheskii Zhurnal, 33 (1): 14–15. (1954). [Preliminary translation by W.E. Ricker]. Fish. Res. Bd. Can., Translation Series No. 177: 17 pp.
31		1958	Laws governing the fluctuations in abundance of important commercial fishes and methods of making catch predictions. By T.F. Dementeva. from Trudy Soveshchanii, 1: 19–36 (1953) [Trudy Konferentsii po Voprosam Rybnovo Khoziaistva, 1951]. Akad. Nauk SSSR, Ikhtiologicheskaia Kommissiia, Moscow. [Preliminary translation by W.E. †Ricker]. Fish. Res. Bd. Can., Translation Series No. 185: 22 pp.
32		1958	Biologically uniform groups among Pacific salmon. By M.L. Krykhtin. from Voprosy Ikhtiologii, No. 10: 3–11. (1958) [Preliminary translation by W.E. Ricker]. Fish. Res. Bd. Can., Translation Series No. 195: 11 pp.
33	I	1959	Vernal and hiemal races among anadromous fishes. By L.S. Berg [translated by W.E. Ricker]. Journal of the Fisheries Research Board of Canada, 16(4): 515–537.
34		1959	Concerning Ricker's paper "Stock and Recruitment". By Yu. E. Lapin from Voprosy Teorii Rybolovstva, 42 pp., (multigraphed), VNIRO, Moscow, (1958). [Preliminary translation by W.E. Ricker (pp. 3–4 and 24–42)]. Fish. Res. Bd. Can., Translation Series No. 204: 24 pp.
35		1959	An open letter to Docent N.N. Andreev of the Department of Commercial Fisheries of the A.I. Mikoian Technical College of Fisheries in Moscow. By P. Moiseev. from Rybnoe Khoziaistvo, No. 2: 51–53 (1955) [Translated by W.E. Ricker]. Fish. Res. Bd. Can., Translation Series No. 211: 5 pp.
36		1959	Biological groups among the Kura osëtr (*Acipenser goldenstaedti persicus* Borodin) and the basis for their artificial culture. By N.L. Gerbilsky. from Doklady Akademii Nauk SSSR, 71(4): 785–788 (1950) [Preliminary translation by W.E. Ricker]. Fish. Res. Bd. Can., Translation Series No. 212: 4 pp.
37		1959	Concerning N.N. Andreev's open letter. By T.F. Dementeva from Rybnoe Khoziaistvo for 1995, No. 2: 53–55 [Preliminary translation by W.E. Ricker]. Fish. Res. Bd. Can., Translation Series No. 213: 4 pp.
38		1959	Intraspecific biological groups among Acipenseridae and the importance of knowledge concerning them for the development of sturgeon culture in connection with hydroelectric developments. By N.L. Gerbilsky from Akademiia Nauk SSSR, Ikhtiologicheskaia Komissiia, Trudy Soveshchanii, 1: 291–300 (1953). [Trudy Konferentsii po Voprosam Rybnovo Khoziaistva, December 17–26, 1951] Moscow. [Preliminary translation by W.E. Ricker]. Fish. Res. Bd. Can., Translation Series No. 233: 11, 1 pp.
39		1960	The American system of fishery management. By F.I. Baranov. from Rybnoe Khoziaistvo for 1946, No. 12, pp. 31–34, (1946). [Preliminary translation by W.E. Ricker]. Fish. Res. Bd. Can., Translation Series No. 70: 7 pp.
41		1960	Instructions for collecting and examining materials obtained in investigations of the food of plankton-eating fishes. By V.G. Bogorov. from Vsesoiuznovo Nauchno-Issledovatelskovo Instituta Morskovo Rybnovo Khoziaistvo i Okeanografii, 15 pp., (1934) [Preliminary translation by W.E. Ricker]. Fish. Res. Bd. Can., Translation Series No. 254: 13 pp.

KR	WER	DATE	
42		1960	Distribution of the young of certain species of fish in temperature gradients. By I.I. Mantelman. from Izvestiia Tikhockeanskovo Nauchno-issledovatelskovo Instituta Rybnovo Khoziaistva, 47 (1): 1–63, (1958) [Translated by the Bureau for Translations, Foreign Language Division, Department of the Secretary of State of Canada, edited by W.E. Ricker]. Fish. Res. Bd. Can., Translation Series No. 257: 58, 9 pp.
44		1960	The question of the migratory impulse in connection with the analysis of intraspecific biological groups. By N.L. Gerbilsky. from Akademiia Nauk SSSR, Ikhtiologicheskaia Komissiia, Trudy Soveshchanii, No. 8: 142–152, (1958). [Trudy Soveshchaniia po Fiziologii Ryb.]. [Preliminary translation. By W.E. Ricker]. Fish. Res. Bd. Can., Translation Series No. 260: 10 pp.
45		1960	The Kuro-sio and the abundance of Amur autumn chum salmon [*Oncorhynchus keta* (Walb.) infrasp. autumnalis]. By I.B. Birman. from Voprosy Ikhtiologii, No. 8: 3–7 (1957) [Preliminary translation by W.E. †Ricker]. Fish. Res. Bd. Can., Translation Series No. 265: 6, 2 pp.
46		1960	Differences in the biology of reproduction and development of residual or dwarf sockeye and anadromous sockeye [Oncorhynchus nerka (Walbaum)]. By A.I. Smirnov from Nauchnye Doklady Vysshei Shkoly, Biologicheskie Nauki, No. 3: 59–65 (1959) [Preliminary translation by W.E. Ricker]. Fish. Res. Bd. Can., Translation Series No. 266: 8, 2 pp.
47		1960	Causes of the decrease in abundance of Sakhalin pink salmon in 1958. By P.A. Dvinin. from Rybnoe Khoziaistvo, 35 (8): 8–9, (1959). [Preliminary translation by W.E. Ricker]. Fish. Res. Bd. Can., Translation Series No. 267: 2 pp.
48		1960	Hydro-electric constructions and fish-passing facilities. By S.M. Kipper. from Rybnoe Khoziaistvo, 35(6): 15–22 (1959). [Preliminary translation by W.E. Ricker]. (Illustrations not included). Fish. Res. Bd. Can., Translation Series No. 268: 8 pp.
49		1960	More about the influence of the Kuro-sio on the dynamics of abundance of salmon. By I.B. Birman from Voprosy Ikhtiologii, (13): 16–18 (1959). [Preliminary translation by W.E. Ricker]. Fish. Res. Bd. Can., Translation Series No. 269: 31 pp.
50		1960	An experiment in rearing the white Amur [Chinese grass carp *Ctenopharyngodon idella*] in ponds of the Angelinsk Fish Rearing Station. By A.P. Lukanin. from Rybnoe Khoziaistvo, 35(6): 82–83 (1959) [Preliminary translation by W.E. Ricker]. Fish. Res. Bd. Can., Translation Series No. 270: 2 pp.
51		1960	The effectiveness of the work of fish-cultural establishments in the Sakhalin region. By I.K. Cherniavskaia. from Rybnoe Khoziaistvo, 35(9): 22–24 (1959) [Preliminary translation by W.E. Ricker]. Fish. Res. Bd. Can., Translation Series No. 271: 3 p.
52		1960	An important problem. By A.A. Klykov. from Nauka i Zhizn for 1958, (1): 79. Moscow. [Preliminary translation by W.E. Ricker]. Fish. Res. Bd. Can., Translation Series No. 272: 2 pp.
53		1960	Effect of intensity of fishing on the stocks of Caspian Shad [Puzanok *Alosa caspia caspia* Svetovidov]. By E.N. Kazancheev. from Rybnoe Khoziaistvo, 35 (4): 16–18 (1959) [Preliminary translation. By W.E. Ricker]. Fish. Res. Bd. Can., Translation Series No. 273: 3, 2 pp.
54		1960	The damage done by lampreys to fish stocks. By V.A. Abakumov. from Rybnoe Khoziaistvo, 35(4): 32–33 (1959) [Preliminary translation by W.E. Ricker]. Fish. Res. Bd. Can., Translation Series No. 274: 2 pp.
55		1960	A study of calcium metabolism with a view to utilizing Ca^{45} as a mark for fish. By M.P. Bogoiavlenskaia. from Vsesoiuznyi N.-I. Institut Morskovo Rybnovo Khoziaistvo i Okeanografii, (VNIRO), 55 pp. (1959) [Preliminary translation of the "Conclusions" only by W.E. Ricker] Fish. Res. Bd. Can., Translation Series No. 276: 2 pp.
56		1960	Some results of mass marking of "non-standard" sturgeon fingerlings. By means of radioactive phosphorus. By G.S. Karzinkin, E.V. Soldatova and I.A. Shekhanova. from Migratsii Zhivotnikh [Animal Migrations], No. 1: 27–40, Akademiia Nauk SSSR. (1959). [Preliminary translation by W.E. Ricker]. Fish. Res. Bd. Can., Translation Series No. 279: 14 pp.
57		1961	Concerning the influence of exploitation on the structure of the population of a commercial fish. By G.V. Nikolsky. from Zoologicheskii Zhurnal, (1): 41–56 (1958) [Preliminary translation by W.E. Ricker]. Fish. Res. Bd. Can., Translation Series No. 280: 19, 6 pp.
58		1960	A method of determining the age and the composition of the catches of the spiny dogfish (*Squalus acanthias* L.). By S. Kaganovskaia. from Vestnik Dalnevostochnovo Filiala Akademii Nauk SSSR for 1933, No. 1–3: 139–141. (1933). Vladivostok. [Also issued as one of a collection of six articles from the above periodical, by the Tikhookeanskii Nauchnyi Institut Rybnovo Khoziaistva, now TINRO] [Preliminary translation by W.E. Ricker]. Fish. Res. Bd. Can., Translation Series No. 281: 3 pp.

KR	WER	DATE	
59		1960	Commercial fishes of the Anadyr River and the Anadyr Estuary. By A.G. Kaganovsky. from Vestnik Dalnevostochnovo Filiala Akademii Nauk SSSR for 1933, No. 1–3: 137–139. (1933). Vladivostok. [Also issued as one of a collection of six articles from the above periodical, by the Tikhookeanskii Nauchnyi Institut Rybnovo Khoziaistva. now TINRO.]. [Preliminary translation by W.E. Ricker]. Fish. Res. Bd. Can., Translation Series No. 282: 5 pp.
60		1960	The parasite fauna and the species relationships of the Kamchatka azabach, *Oncorhynchus nerka* infrasp. *asabatch* Berg (1932). By A. Kh. Akhmerov. from Doklady Akademii Nauk SSSR, 94 (5): 969–971. (1954). [Preliminary translation by W.E. Ricker]. Fish. Res. Bd. Can., Translation Series No. 283: 4 pp.
61		1960	Local stocks of pink salmon in the Amur basin and neighbouring waters. By R.I. Eniutina. from Voprosy Ikhtiologii, No. 2: 139–143 (1954) [Preliminary translation by W.E. Ricker]. Fish. Res. Bd. Can., Translation Series No. 284: 8 pp.
62		1960	Lake coho, *Oncorhynchus kisutch* (Walbaum) morpha *relictus* nova. By P.A. Dvinin. from Doklady Akademii Nauk SSSR, 69(5): 695–697 (1949) [Preliminary translation by W.E. Ricker]. Fish. Res. Bd. Can., Translation Series No. 285: 3 pp.
63		1960	The biology and production of the stocks of Black Sea kumzha (salmon/trout). By G.P. Barach. from Akademiia Nauk SSSR, Ikhtiologicheskaia Komissiia, Trudy Soveshchanii, (7): 235–242 (1957) [Trudy Soveshchaniia po Rybovodstvu, 1954]. [Preliminary translation by W.E. Ricker]. Fish. Res. Bd. Can., Translation Series No. 285: 10 pp.
64		1960	Morphobiological and morphometric characteristics of the pink salmon of the Amgun and Iski Rivers. By R. Eniutina. from Izvestiia Tikhockeanskii N.-I. Institut Rybnovo Khoziaistva i Okeanografii, 41: 333–336 (1954) [Preliminary translation by W.E. Ricker]. Fish. Res. Bd. Can., Translation Series No. 289: 7 pp.
65		1960	Parasitism of salmon of the genus *Oncorhynchus* by the Pacific lamprey. By I.B. Birman. from Izvestiia TINRO, (32): 158–160 (1950) [Preliminary translation by W.E. Ricker]. Fish. Res. Bd. Can., Translation Series No. 290: 3 pp.
66		1960	Pink salmon in the basins of the Barents and White Seas. By E.G. Kossov, M.S. Lazarev & L.V. Polikashin. from Rybnoe Khoziaistvo, 36(8): 20–25 (1960) [Preliminary translation by W.E. Ricker]. Fish. Res. Bd. Can., Translation Series No. 323: 9 pp.
67		1961	Bering Sea halibuts. By N.P. Novikov. from Rybnoe Khoziaistvo, 36(1): 12–15 (1960) [Translation by H. Kasahara and W.E. Ricker]. Fish. Res. Bd. Can., Translation Series No. 329: 4 pp. (Translation No. 9 of the International North Pacific Fisheries Commission)
68		1961	Availability of sticklebacks as food for the predacious fishes of the Volga Delta. By K.P. Fortunatova. from Zoologicheskii Zhurnal, 38(11): 1689–1701 (1959) [Translation by John Magnuson, edited by W.E. Ricker]. Fish. Res. Bd. Can., Translation Series No. 331: 18 pp.
69		1961	Zoogeographical analysis and a hypothesis of the origin of the helminth fauna of the vertebrates of Sakhalin. By A.I. Krotov. from Zoologicheskii Zhurnal, 39(4): 481–489 (1960) [Preliminary translation by W.E. Ricker]. Fish. Res. Bd. Can., Translation Series No. 334: 12 pp.
70		1961	The importance of fish diseases in the production of fish stocks. By V.A. Dogel. from Ikhtiologicheskaia Komissiia AN SSSR, Trudy Soveshchanii, (7): 117–123 (1957) [Preliminary translation by W.E. Ricker, Bibliography not included]. Fish. Res. Bd. Can., Translation Series No. 343: 15 pp.
71		1961	Local stocks of summer keta, *Oncorhynchus keta* (Walbaum), of the Amur basin. By A.A. Svetovidova. from Voprosy Ikhtiologii, 17: 14–23 (1961) [Preliminary translation by W.E. Ricker] Fish. Res. Bd. Can., Translation Series No. 347: 16 pp.
72		1961	Local stocks of autumn chum salmon in the Amur basin. By I.B. Birman. from Voprosy Ikhtiologii, 7: 158–173 (1956) [Preliminary translation by W.E. Ricker]. Fish. Res. Bd. Can., Translation Series No. 349: 22 pp.
73		1961	Distinctive features of the biology of the masu salmon (*Oncoryhnchus masu* Brevoort) of Sakhalin by P.A. Dvinin. from Voprosy Ikhtiologii, (7): 33–35 (1956) [Preliminary translation. By W.E. Ricker]. Fish. Res. Bd. Can., Translation Series No. 350: 5 pp.
74		1961	Concerning the acclimatization of the grass carp [*Ctenopharyngodon*] in the lower reaches of the Volga River. By N.P. Tanasiichuk. from Voprosy Ikhtiologii (17): 176–178 (1961) [Preliminary translation by W.E. Ricker]. Fish. Res. Bd. Can., Translation Series No. 352: 5 pp.

KR	WER	DATE	
75		1961	Seasonal races among anadromous fishes. By V.A. Abakumov. from Voprosy Ikhtiologii, (17): 179–190 (1961) [Preliminary translation by W.E. Ricker]. Fish. Res. Bd. Can., Translation Series No. 353: 18 pp.
76		1961	The effect of the fishery on the productivity of bream stocks of the Aral Sea. By E.A. Bervald. from Voprosy Ikhtiologii, (7): 21–32 (1956) [Preliminary translation by W.E. Ricker]. Fish. Res. Bd. Can., Translation Series No. 354: 15 pp.
77		1961	Information on the ecology of rockfishes (family Scorpaenidae) of the northern part of the Pacific Ocean. By P.A. Moiseev & I.A. Paraketsov. from Voprosy Ikhtiologii, 1 (1–18): 39–45 (1961) Moscow. [Preliminary translation by W.E. Ricker]. Fish. Res. Bd. Can., Translation Series No. 358: 10 pp.
78		1961	Arctic fish culture. Anonymous. from Rybnoe Khoziaistvo, 37 (4): 42–47. (1961). [Preliminary translation by W.E. Ricker]. Fish. Res. Bd. Can., Translation Series No. 359: 3 pp.
79		1961	Acclimatizaton of Pacific salmon in the Barents and White Seas. By A.I. Isaev. from Voprosy Ikhtiologii, 1 (1–18): 46–51. (1961). Moscow. [Preliminary translation by W.E. Ricker]. Fish. Res. Bd. Can., Translation Series No. 361: 7 pp.
81		1961	Results of three years of observations on fish below the Lenin Hydroelectric Station on the Volga. By A.T. Diuzhikov. from Voprosy Ikhtiologii, 1 (1–18): 69–78 (1961) Moscow [Preliminary translation by W.E. Ricker]. Fish. Res. Bd. Can., Translation Series No. 363: 16 pp.
82		1961	An analysis of the mechanism of distribution of fish in a temperature gradient. By V.S. Ivlev. from Zoologicheskii Zhurnal, 39(4): 494–499 (1960) [Preliminary translation by W.E. Ricker]. Fish. Res. Bd. Can., Translation Series No. 364: 8 pp.
83		1962	The formation of the thermocline in lakes. By E.M. Krokhin. from Izvestiia Akademii Nauk SSSR, Seriia geograficheskaia, (6): 90–97 (1960) [Translation by R.E. Foerster; edited by W.E. Ricker]. Fish. Res. Bd. Can., Translation Series No. 365: 6, 2 pp.
84		1961	A method of relative evaluation of the condition of a stock and prediction of expected catch of fishes having a short life cycle. By Liudiia St. Ivanov. from Voprosy Ikhtiologii, (14): 160–165. (1960). [Preliminary translation by W.E. Ricker]. Fish. Res. Bd. Can., Translation Series No. 366: 8 pp.
85		1962	The coho salmon [*Oncorhynchus kisutch* (Walbaum)] a biological sketch. By V.I. Gribanov. from Izvestiia Tikhookeanskogo N.-I. Instituta Rybnogo Khoziaistva i Okeanografii, (TINRO), 28: 43–101 (1948) [Preliminary translation by W.E. Ricker]. Fish. Res. Bd. Can., Translation Series No. 370: 83 pp.
87		1962	On the mechanisms of adaptation to changes in water salinity by sockeye salmon [*Oncorhynchus nerka* (Walb.)]. By M.G. Zaks and M.M. Sokolova. from Voprosy Ikhtiologii, 1(2): 333–346 (1961) [Translated. By R.E. Foerster; edited by W.E. Vanstone and W.E. Ricker]. Fish. Res. Bd. Can., Translation Series No. 372: 13, 5 pp.
88		1963	First results of the operation of the fish pass at the Volga Hydroelectric Station named "XXII Congress of the KPSS". By L.M. Nusenbaum and L.N. Lapitskaya. from Rybnoe Khoziaistvo, 37(11): 13–18 (1961) [Preliminary translation by W.E. Ricker; edited by C.H. Clay]. Fish. Res. Bd. Can., Translation Series No. 378: 4 pp.
89		1962	Seasonal changes in the blood composition of the river lamprey (*Lampetra fluviatilis* L.). By M.M. Sokolova. from Voprosy Ikhtiologii, (13): 156–162 (1959) [Preliminary translation by W.E. Ricker]. Fish. Res. Bd. Can., Translation Series No. 383: 13 pp.
90		1962	Concerning the causes of fluctuations in the abundance of fishes. By G.V. Nikolsky. from Voprosy Ikhtiologii, 1 (4–21): 659–665 (1961) [Preliminary translation by W.E. Ricker]. Fish. Res. Bd. Can., Translation Series No. 389: 11 pp.
91		n.d.	The origin of glacial relict amphipods in relation to the question of the postglacial connection between the White Sea and the Baltic. By N.B. Lomakina. from Uchenye Zapiski Karelo-Finskovo Universiteta, Biologicheskie Nauki, IV (3): 110–127 (1952) [Preliminary translation by W.E. Ricker]. Fish. Res. Bd. Can., Translation Series No. 390: 17 pp.
92		1962	Parasitic diseases of young Pacific salmon in fish hatcheries of the Far East. By E.A. Bogdanova. from Nauchno-Tekhnicheskii Biulleten GosNIORKh, (11): 49–52 (1960) [Preliminary translation by W.E. Ricker]. Fish. Res. Bd. Can., Translation Series No. 391: 4, 1 pp.
95		1962	The biological productivity of waters. By V.S. Ivlev. from Uspekhi Sovremennoi Biologii (Advances in Modern Biology), 19 (1): 98–120 (1945) [Preliminary translation by W.E. Ricker (1946)]. Fish. Res. Bd. Can., Translation Series No. 394: 28 pp.

KR	WER	DATE	
96		1962	Certain biological and practical conclusions from the experiments on transplantation of Pacific salmon. By A.I. Smirnov. from Biulleten Moskovskogo Obshchestva Ispytatelei Prirody, otd. Biologii, 67 (1): 149–151 (1962) [Translated by R.E. Foerster and W.E. Ricker]. Fish. Res. Bd. Can., Translation Series No. 408: 5 pp.
97		1963	Contemporary changes in the abundance of the Pacific salmons, and their principal causes. By R.S. Semko. from Trudy Soveshchanii Ikhtiologicheskoi Komissii Akademii Nauk SSSR, 13: 117–129 (1961) [Soveshchanie po Dinamike Chislennosti Ryb 1960]. [Preliminary translation by W.E. Ricker]. Fish. Res. Bd. Can., Translation Series No. 412: 15, 3 pp.
98		1964	The rate of growth and age groupings of sockeye salmon (*Oncorhynchus nerka* Walb.) in the sea. By F.V. Krogius. from Voprosy Ikhtiologii, 16: 67–88. (1960). [Translated by R.E. Foerster and W.E. Ricker]. Fish. Res. Bd. Can., Translation Series No. 413: 23, 5 pp.
99		1963	Basic factors in the fluctuations of the stocks of sturgeons and lesser fishes in the Sea of Azov. By E.G. Boiko. from Trudy Soveshchanii Ikhtiologicheskoi Komissii Akademii Nauk SSSR, 13: 147–157 (1961) [Soveshchanie po Dinamike Chislennosti Ryb 1960]. [Preliminary translation by W.E. Ricker]. Fish. Res. Bd. Can., Translation Series No. 414: 9, 4 pp.
100		1963	Primary production in the Atlantic ocean. By Yu. I. Sorokin & L.B. Kliashtorin. from Trudy Vsesoiuznogo Gidrobiologicheskogo Obshchestva, 11: 265–284 (1961). [Preliminary translation by W.E. Ricker; edited by J.D.H. Strickland]. Fish. Res. Bd. Can., Translation Series No. 421: 18, 5† pp.
101		1963	Physiological adaptations of fishes to new conditions of existence. By T.I. Privolnev. from Trudy Soveshchanii Ikhtiologicheskoi Komissii Akademii Nauk SSSR, 3: 40–49 (1954) [Trudy Soveshchania po probleme akklimatizatsii ryb i kormovykh bespozvonochnykh, 1952]. [Preliminary translation by W.E. Ricker]. Fish. Res. Bd. Can., Translation Series No. 422: 10 pp.
103		1963	The energy principle in studying food associations and the productivity of ecological systems. By G.G. Winberg. from Zoologicheskii Zhurnal, 41(11): 1618–1630 (1962) [Preliminary translation by W.E. Ricker]. Fish. Res. Bd. Can., Translation Series No. 433: 14, 3 pp.
104		1963	On the principles of fish population dynamics. By T.F. Dementeva, Yu. Yu. Marti, P.A. Moiseev & G.V. Nikolsky. from Trudy Soveshchanii Ikhtiologicheskoi Komissii Akademii Nauk SSSR, 13: 7–20 (1961) [Soveshchanie po Dinamike Chislennosti Ryb, 1960]. [Preliminary translation by W.E. Ricker]. Fish. Res. Bd. Can., Translation Series No. 435: 25 pp.
105		1963	Information on the biology of pink salmon acclimatized in the basins of the White and Barents Seas. By V.V. Azbelev & S.S. Surkov. from Nauchno-tekhnicheskii biulleten PINRO, 2–3 (20–21): 37–38 (1962) [Preliminary translation by W.E. Ricker]. Fish. Res. Bd. Can., Translation Series No. 437: 3 pp.
106		1963	On the effect of feeding conditions and abundance of spawners on the amount of progeny produced. By Caspian bream. By K.A. Zemskaia. from Trudy Soveshchanii Ikhtiologicheskoi Komissii Akademii Nauk SSSR, 13: 307–313 (1961) [Soveshchanie po Dinamike Chislennosti Ryb, 1960]. [Preliminary translation by W.E. Ricker]. Fish. Res. Bd. Can., Translation Series No. 438: 6, 4 pp.
107		1963	The trawl fishery for rockfish in the Bering Sea. By A.V. Lestev. from Sakhalinskoe Otdeleinie Tikhookeanskogo N.-I. Instituta Morskogo Rybnogo Khoziaistva i Okeanografii (TINRO), "Tekhnika Rybolovsta" Series, 26 pp. Rybnoe Khoziaistvo Press, Moscow (1961) [Preliminary translation by W.E. Ricker]. Fish. Res. Bd. Can., Translation Series No. 439: 13 pp.
108		1963	Information on the biology and survival of salmon of the Kola Peninsula. By V.V. Azbelev. from Trudy Poliarnogo n.-i. Instituta Morskogo Rybnogo Khoziaistva i Okeanografii imeni N.M. Knipovich (PINRO), 12: 5–70 (1960) [Preliminary translation of pp. 5–9, 19–27 and 51–65 by W.E. Ricker; tables and figures are omitted]. Fish. Res. Bd. Can., Translation Series No. 440: 19 pp.
109		1963	Rockfish in the Gulf of Alaska. By T.G. Liubimova. from Rybnoe Khoziaistvo, 37(9): 27–30 (1961) [Preliminary translation by W.E.† Ricker]. Fish. Res. Bd. Can., Translation Series No. 441: 2, 2 pp.
110		1963	Plankton and the food of larval Baltic herring, in the Gulf of Riga. By L.N. Lisivnenko. from Trudy N.-I Instituta Rybnoe Khoziaistva Soveta Narodnogo Khoziaistva Latviiskoi SSR (LatvNIRO), 3: 105–138 (1961) Riga. [Preliminary translation by W.E. Ricker]. Fish. Res. Bd. Can., Translation Series No. 444: 34, 2 pp.
111		1963	Catching kilka with a fish pump using underwater illumination and a pulsing electric current. By I.V. Nikonorov & A. Kh. Pateev. from Rybnoe Khoziaistvo, 35: 53–58 (1959) [Preliminary translation by W.E.† Ricker]. Fish. Res. Bd. Can., Translation Series No. 445: 5, 4 pp.
112		1963	The use of blocks of lights in fishing for kilka with a fish pump. By I.V. Nikonorov. from Rybnoe Khoziaistvo 35(12): 42–47 (1959) [Preliminary translation by W.E. Ricker]. Fish. Res. Bd. Can., Translation Series No. 446: 5, 2 pp.

KR	WER	DATE	
113		1963	On the utilization of food. By plankton-eating fishes. By V.S. Ivlev. from Trudy Sevastopoloskoi Biologicheskoi Stantsii, 14: 188–201 (1961) [Preliminary translation by W.E. Ricker]. Fish. Res. Bd. Can., Translation Series No. 447: 15, 1 pp.
114		1963	A pink salmon from a lake. By E.L. Bakshtansky. from Biulletin PINRO, 4 (22): 46–47 (1962) Murmansk. [Preliminary translation by W.E. Ricker]. Fish. Res. Bd. Can., Translation Series No. 451: 3 pp.
115		1964	Changes in weight and length of fish after fixation in formalin followed by soaking in water. By V.A. Amosov. from Voprosy Ikhtiologii, 16: 187–190 (1960) [Preliminary translation by W.E. Ricker]. Fish. Res. Bd. Can., Translation Series No. 452: 22 pp.
117		1964	Concerning the supplementary ring in the nuclear portion of the scales of pink salmon *Oncorhynchus gorbuscha* (Walb.). By R.I. Eniutina. from Voprosy Ikhtiologii, 2(4): 740–742 (1962) [Old Series No. 25]. [Preliminary translation by W.E. Ricker]. Fish. Res. Bd. Can., Translation Series No. 510: 7 pp.
121		1964	Evaluating natural mortality in the Azov zander [*Stizostedion lucioperca*]. By E.G. Boiko. from Trudy Vsesoiuznogo Nauchno-issledovatelskogo Instituta Morskogo Rybnoe Khoziaistva i Okeanografii (VNIRO), 50: 143–161 (1964) [Preliminary translation by W.E. Ricker]. Fish. Res. Bd. Can., Translation Series No. 541: 15, 16 pp.
122		1964	The life cycle of the pink salmon of Primor'ye. By N.V. Dubrovskaia. from Rybnoe Khoziaistvo Dal'nego Vostoka, 1–2: 65–68 (1934) [Preliminary translation by W.E. Ricker. Figures not included]. Fish. Res. Bd. Can., Translation Series No. 542: 6 pp.
123		1965	A natural monument or a commercial species? The future of the Commander Islands sea otters. By S.V. Marakov. from Priroda, 52 (11): 79–83 (1963) [Preliminary translation by W.E. Ricker, figures not included.]. Fish. Res. Bd. Can., Translation Series No. 543: 6 pp.
124		1965	New information on within-species differences among the acipenserid fishes of the Caspian Sea. By G.A. Amrikhanov. from Rybnoe Khoziaistvo, 40(8): 15–18 (1964) [Preliminary translation by W.E. Ricker]. Fish. Res. Bd. Can., Translation Series No. 546: 5, 2 pp.
125		1965	Saury fishing with a pump using light and electric current. By I.V. Nikonorov & A. Kh. Pateev. from Rybnoe Khoziaistvo, 39(2): 51–53 (1962) [Preliminary translation by W.E. Ricker]. Fish. Res. Bd. Can., Translation Series No. 547: 5, 1 pp.
126		1965	Was Steller's sea cow exterminated? By A.A. Berzin, E.A. Tikhomirov & V.I. Troinin. from Priroda, 52 (8): 73–75 (1963) [Preliminary translation by W.E. Ricker. Illustration not included]. Fish. Res. Bd. Can., Translation Series No. 548: 4 pp.
127		1965	On the question of the effect of growth during the first years of a fish's life on its subsequent growth. By D.F. Zamakhaev. from Trudy Vsesoiuznogo Nauchno-issledovatel'skogo Instituta Morskogo Rybnogo Khoziaistva i Okeanografii (VNIRO), (50): 109–141 (1964) [Preliminary translation by W.E. Ricker]. Fish. Res. Bd. Can., Translation Series No. 549: 39, 16 pp.
128		1965	The sea otter and stages in its investigation. By I.I. Barabash-Nikiforov. from Priroda for 1938, 2: 51–61 (1938) [Preliminary translation by W.E. Ricker (figures not included)]. Fish. Res. Bd. Can., Translation Series No. 551: 16 pp.
129		1965	In search of the "devil" of Lake Labynkyr. By V.A. Koshelev and V.I. Frumson. from Priroda, 52(3): 83–89 (1963) [Preliminary translation by W.E. Ricker (figures not included)]. Fish. Res. Bd. Can., Translation Series No. 552: 17 pp.
130		1965	Methods of studying the growth of fish from their scales, bones and otoliths. By V.L. Briuzgin. from Voprosy Ikhtiologii, 3(2): 347–j365 (1963) (Old Series No. 27) [Preliminary translation by W.E. Ricker]. Fish. Res. Bd. Can., Translation Series No. 553: 31, 3 pp.
131		1965	Some questions concerning the fishery for deepwater redfish in the Newfoundland region. By V.A. Rikhter. from Rybnoe Khoziaistvo, 41(5): 15–17; and 41 (7): 11–12 (1965) [Preliminary translation by W.E. Ricker]. Fish. Res. Bd. Can., Translation Series No. 555: 6, 2 pp.
134		1965	The natural mortality factor and its importance in regulating fisheries. By P.V. Tiurin. from Voprosy Ikhtiologii, 2(3): 403–427 (1962) (Old Series No. 24). [Preliminary translation by W.E. Ricker]. Fish. Res. Bd. Can., Translation Series No. 558: 49 pp.
135		1966	On the uses of mathematics in ichthyology. By F.I. Baranov. from Rybnoe Khoziaistvo, 41(8): 4–8, and 41(9): 6–8 (1965) [Preliminary translation by W.E. Ricker]. Fish. Res. Bd. Can., Translation Series No. 559: 13 pp.
136		1965	A new contribution to the development of methods for determining the age of baleen whales. By T.V. Andreeva & A.V. Yablokov. from Zoologicheskii Zhurnal, 40(1): 145–146 (1965) [Preliminary translation by W.E. Ricker]. Fish. Res. Bd. Can., Translation Series No. 560: 3, 2 pp.

KR	WER	DATE	
137		1965	On methods of harvesting fur seals. By V.A. Arsen'ev. from Rybnoe Khoziaistvo, 41(9): 55–57 and 41(10): 50–52 (1965) [Preliminary translation by W.E. Ricker]. Fish. Res. Bd. Can., Translation Series No. 700: 6 pp.
138		1965	Model of a fish population with a variable recruitment structure. By V.V. Menshutkin. from Voprosy Ikhtiologii, 5(2): 235–239 (1965) (Old Series No. 35). [Preliminary translation by W.E. Ricker]. Fish. Res. Bd. Can., Translation Series No. 702: 8, 3 pp.
139	II	1966	The biological productivity of waters. By V.S. Ivlev. [Translated by W.E. Ricker]. Journal of the Fisheries Research Board of Canada, 23(11): 1727–1759.
140	III	1965	In memoriam – Alexander Nikolaevich Derzhavin. By Yu. A. Abdurakhmanov & N.I. Chugunova [Translated by W.E. Ricker]. Canadian Fish Culturist, 35: 67–72.
141		1966	Intraspecific variability in respect to oxyphilia mong freshwater fishes. By G.L. Shkorbatov. from Gidrobiologicheskii Zhurnal, 1(5): 3–8 (1965) [Preliminary translation by W.E. Ricker]. Fish. Res. Bd. Can., Translation Series No. 701: 16 pp.
142		1966	On the age of structure of pink salmon populations [*Oncorhynchus gorbuscha* (Walb.)]. By V.N. Ivankov. from Voprosy Ikhtiologii, 5(4): 622–667 (1965) [Old Series No. 37]. [Preliminary translation by W.E. Ricker]. Fish. Res. Bd. Can., Translation Series No. 705: 9, 5 pp.
143		1966	On the size of the commercial return among sturgeons. By A.F. Gun'ko & V.M. Naumov. from Voprosy Ikhtiologii, 5(4): 652–656 (1965) [Old Series No. 37]. [Preliminary translation by W.E. Ricker]. Fish. Res. Bd. Can., Translation Series No. 706: 10 pp.
144		1966	Concerning the most profitable size for fish. By F. Baranov. from Byulleten' Rybnogo Khozyaistva for 1925, No. 11: 32. [Preliminary translation by W.E. Ricker]. Fish. Res. Bd. Can., Translation Series No. 712: 3 pp.
145		1966	Concerning the releasing of young pink and chum salmon from fish hatcheries. By S.P. Volovik. from Rybnoe Khoziaistvo, 42(4): 13–14 (1966) [Preliminary translation by W.E. Ricker]. Fish. Res. Bd. Can., Translation Series No. 713: 4 pp.
146		1966	A new stage in the culture of plant-eating fishes. By K.E. Babayan. from Rybnoe Khoziaistvo, 42(6): 4–7 (1966) [Preliminary translation by W.E.† Ricker]. Fish. Res. Bd. Can., Translation Series No. 714: 8 pp.
147		1966	Measures for the development of our fisheries. By V.A. Kevdin. from Sovremennoe Rybolovstvo Rossii, pp. 112–145. Moscow, (1915). Reprinted in Ocherki po Biologicheskim Osnovam Rybnogo Khozyaistva, pp. 62–68. Akademiya Nauk SSSR, Moscow (1961) [Preliminary translation by W.E. Ricker]. Fish. Res. Bd. Can., Translation Series No. 715: 11 pp.
148		1966	Changes in fatness of young sockeye as related to their downstream migration. By V.N. Akulin. from Rybnoe Khoziaistvo, 42(8): 11–12 (1966) [Preliminary translation by W.E. Ricker]. Fish. Res. Bd. Can., Translation Series No. 716: 4 pp.
149		1967	The magnitude of the primary production of the Pacific Ocean. By O.I. Koblenz-Mishke. from Okeanologiya, 5(2): 325–337 (1965) [Preliminary translation by W.E. Ricker]. Fish. Res. Bd. Can., Translation Series No. 828: 18, 6 pp.
150		1967	The fishing industry and the age limit among fishes. By F. Baranov. from Byulleten' Rybnogo Khoziaistva, No. 9: 26–27 (1925) [Preliminary translation by W.E. Ricker]. Fish. Res. Bd. Can., Translation Series No. 829: 4 pp.
151		1967	Herring of the Korfa-Karagin region. By T.F. Kachina & V.G. Prokhorov. from Rybnoe Khoziaistvo, 42(12): 4–5 (1966) [Preliminary translation by W.E. Ricker]. Fish. Res. Bd. Can., Translation Series No. 830: 5 pp.
152		1967	Downstream migration and behaviour of the young of introduced pink salmon. By M.S. Kamyshnaya. from Rybnoe Khoziaistvo, 43(1): 9–12 (1967) [Preliminary translation by W.E. Ricker]. Fish. Res. Bd. Can., Translation Series No. 833: 8 pp.
153		1967	On the diel rhythm of feeding among Baikal grayling fry, and on diel rhythms among young fish in general. By V.I. Olifan. from Doklady Akademii Nauk SSSR, 114(3): 669–672 (1957) [Preliminary translation by W.E. Ricker]. Fish. Res. Bd. Can., Translation Series No. 834: 9 pp.
154		1967	An experiment to improve conditions for reproduction of the Pacific saury. By B.N. Ayushin, O.P. Kodolova & Yu. V. Novikov. from Rybnoe Khoziaistvo, 43(5): 10–11 (1967) [Preliminary translation by W.E. Ricker]. Fish. Res. Bd. Can., Translation Series No. 835: 3 pp.
155		1967	Photosynthesis. By A.A. Nichiporovich. from Priroda 1967, 6: 33–43 (1967) [Preliminary translation by W.E. Ricker]. Fish. Res. Bd. Can., Translation Series No. 837: 19, 6 pp.

KR	WER	DATE	
156		1967	On the efficiency of utilization of energy in pelagic ecosystems of the Black Sea. By T.S. Petipa. from Struktura i dinamika vodnykh soobshchestv i populyatsii, pp. 44–64 (1967) Respublikanskii Mezhvedomstvennyi Sbornik, Seriya "Biologiya Morya". Adakemiya Nauk Ukrainskoi SSR. Published by "Naukova Dumka", Kiev. [Preliminary translation by W.E. Ricker]. Fish. Res. Bd. Can., Translation Series No. 973: 34 pp.
157		1968	Productive regions of the ocean. By V.G. Bogorov. from Priroda 1967, 10: 40–46 [Preliminary translation by W.E. Ricker]. Fish. Res. Bd. Can., Translation Series No. 985: 17 pp.
172		1971	On the problems of estimating the supply of zanders in the North Caspian. By I.N. Voevodin. from Trudy Pervoi Vsekaspiiskoi Nauchnoi Rybokhozyaistrennoi Konferentsii, 2: 3–12 (1938) (in Russian). [Preliminary translation by W.E. Ricker]. Fish. Res. Bd. Can., Translation Series No. 2153: 9 pp.

*All references that have been omitted from this Compilation are available from the Pacific Biological Station, Nanaimo, British Columbia, Canada.

Selected translations from Russian, not published (with no dates) (in W.E. Ricker personal and archival files)

1. More about the poor catch of vobla. F.I. Baranov from Byull. R'ybn. Khoz. 7(7): pp? (1927). (in Russian) [KR No. 173].

2. An attempt at a biostatistical determination of the stocks of fishes in the North Caspian. N.L. Chugunov. Rybn. Khoz. 15(6): 24–29; 15(8): 17–12. (1935) (in Russian) [KR No. 184].

3. A method of making long term forecasts for the vobla fishery of the North Caspian. G.N. Monastrysky. Rybn. Khoz. 15(5): 11–18; 15(6); 18–23. (1935) (in Russian) [KR No. 183].

4. Developmental stages of bony fishes. V.V. Vasnetsov. pp. 207–217, in Ocherki po Obshchim Voprosam Ikhtiologii (1938). Akademiya Nauk Press, Moscow (in Russian) [KR No. 230].

5. Notes to the third volume of F.I. Baranov's collected works. A. Zasosov. pp. 268–302, in Pishchevaya Promyshlennost Press, Moscow (1971) (in Russian) [KR No. 231].

Selected translations from Russian

6. James Cook's clock. Leonid Pasenyuk. from booklet Nord Ost (1980) pp. 128–136 [Published by Dalnevostochnoe Krishnoe Izdatelstvo, Kamchatskoe Otdelnie, Petropavlovsk-Kamchatskii]. [Translation by W.E. Ricker and Z. Kabata]. 15 pp + 2 pp notes [KR No. 228].

7. The chatelaine of Lake Dalnee. G. Vasileva. from booklet Nord Ost (1980) pp. 3–12 [Published, ibid]. [Translation by W.E. Ricker]. 18 pp [KR No. 232].

8. Was Frederick Cook at the North Pole? V.S. Koryakin. Priroda, 7(719): 74–84 (1975) [Translation by W.E. Ricker] 16 pp [KR No. 233].

9. The Pacific sockeye salmon in the ecosystem of Lake Dalnee. F.V. Krogius, E.M. Krokhin & V.V. Menshutkin. Nauka Press (Leningrad Div.): 200 pp. (1997) [Translation by W.E.Ricker]. Excerpts, beginning on page 107. 40 pp [KR No. 234].

There are 54 additional unpublished translations, plus one German translation. Contact Karl Ricker for titles and photocopies.

Selected unpublished manuscripts (some on file at library, Pacific Biological Station, Nanaimo, B.C., not in chronologic order). 36 titles out of total of 173 are listed. Contact Karl Ricker for a complete listing and photocopies.

c		
i	1944	Inflation. MS: 23 pp. (Lecture presented to Indiana University's "Professor's Club").
ii	1957	Estimation of potential sustained yield of salmon from the Fraser River and tributaries above Yale, B.C. 46 pp.
iii	1959	Stoneflies from the Rocky Mountain Parks, 4 pp.
iv	1960	Report on visit to Moscow and Leningrad, 28 January to 12 February 1960. 41 pp.

v	1961	Report on visit to research laboratories dealing with fishes and marine mammals in Tokyo (6–8 February 1961), 5 pp.
vi	1967	Personal diary of a visit to the USSR and Norway, 25 March–9 April 1967. 9 pp.
vii	1970	Report on the Symposium "Stock and Recruitment", Aarhus, Denmark, 7–10 July 1970. 11 pp. + 3 pp. Appendix.
viii	1971	Comments on the West Atlantic harp seal herd and proposals for the 1972 harvest. 25 pp. (reproduced as an appendix to publication No. 172a (WER #).
ix	1972	Biological aspects of fishery management. 13 pp.
x	1972	Comments on fur seal studies. 13 pp.
xi	1973	Dr. R.E. Foerster had active career. Nanaimo Free Press, 19 September 1973.
xii	1980	Notes on certain of the testimonial documents that pertain to the effects of power plants on striped bass in the lower Hudson River and estuary. 46 pp., 7 figures.
xiii	1980	Comments on a report. By W.E. Johnson, concerning salmon production. 11 pp. + 4 rough figures.
xiv	Various dates	Historic routes in the Fraser River canyon. (approximate title) (several drafts circulated in the 1980s and early 1990s). ca 300 pp.
xv	1989	Words used in B.C. (re Chinook). Letter to T. Parker, Nanaimo, 15 October 1989.
xvi	1993	The great meteoric display of 9 February 1913. (published, see WER #197), also see #210 (1999) in main list of publications.
xvii	1995	The role of Pacific salmon in their ocean environment. MS: 15 pp.
xviii	1978	Notes on the research and information needs for the British Columbia salmon fisheries. MS: 6 pp.
xix	1982	Impact of "enhanced" populations on natural stocks. MS: 3 pp.
xxb	1948	New evidence on the passability of Hell's Gate to salmon, 1915–1944. Indiana University, MS: 42 pp.
xxa	1955–58	Rehabilitation of the Fraser sockeye. MS: 12 pp.
xxii	1954	Memorandum on introduction of Pacific salmon into tributaries of Hudson and James Bay. MS: 3 pp.
xxiii	1954	Report on an inspection of the Winisk and Attawapiskat rivers and adjacent waters, from the point of view of their suitability for salmon propagation, 19–26 July 1954. MS: 23 pp + 2 maps.
xxiv	1962	Further information on the qualification of Canadian herring stocks for abstention under Article IV 1(b) (i). F.H.C. Taylor & W.E. Ricker. MS: 16 †pp.
xxv	1960	Notes on some trends in fishery investigation and theory in the USSR. MS: 11 †pp.
xxvi	1937	A synopsis of the investigations at Cultus Lake, British Columbia, conducted by the Biological Board of Canada into the life history and propagation of sockeye salmon, 1924–1937. R.E. Foerster & W.E. Ricker. MS: 21 pp, December, 1937.
xxvii	1955	Some consequences of the theory of random encounters in determining the density of and yield from an animal population, in Lectures on Population Dynamics. University of California at San Diego (Scripps Institute of Oceanography). MS:† pp. 35–41 + 3 figures.
xxviii	1949	The lake fishes of northern Indiana. MS: 9 pp. (part of a much longer document.)
xxix	1966	Final report on FRB library services and indexing, and on the availability of scientific information. MS: 28 pp.
xxx	1965	Report on a visit to the Aquatic Sciences Information Retrieval Center (ASIRC). University of Rhode Island, 14–15 January 1965. MS: 4 pp.
xxxi	1965	Interim report on two annual indexes for fishery literature. MS: 4 pp.
xxxii	1940	Some statistical procedures for fisheries biologists. William E. Ricker & Albert L. Tester. MS (partly in long hand): 49 pp.
xxxiii	1945	Analysis of Dr. Mottley's data on size of trout caught in Paul Lake. MS: 6 pp + 4 tables.
xxxiv	1983	Interaction between coho and pink salmon. (*Oncorhynchus kisutch* and *O. gorbuscha*). MS: 11 pp.
xxxv	1947	Probability and likelihood. Lecture presented to Indiana University's "Professor's Club", MS: ca 10 pp.
xxxvi	1971	Marine fish production, yield potential and management problems, 11 pp + table and 8 figures on attachment.
lxxxvb	2005	[Ricker, W. E., E.G. Pogodaev, R. Beamish and V. I. Karpenko] cases of rapid changes in rate of freshwater growth and ages of smolting and maturation among sockeye salmon of a small Kamchatka Lake (to be published in Russian).

GPSR Compliance

The European Union's (EU) General Product Safety Regulation (GPSR) is a set of rules that requires consumer products to be safe and our obligations to ensure this.

If you have any concerns about our products, you can contact us on ProductSafety@springernature.com

In case Publisher is established outside the EU, the EU authorized representative is:

Springer Nature Customer Service Center GmbH
Europaplatz 3
69115 Heidelberg, Germany

Batch number: 10615455

Printed by Printforce, the Netherlands